Lecture Notes in Operations Research and Mathematical Systems

Economics, Computer Science, Information and Control

56

Mohamed A. El-Hodiri
University of Kansas, Department of Economics,
Lawrence, Kansas/USA

Constrained Extrema Introduction to the Differentiable Case with Economic Applications

Springer-Verlag
Berlin · Heidelberg · New York 1971

AMS Subject Classifications (1970): 34 H 05, 49 B xx, 90 A 99

ISBN-13: 978-3-540-05637-9 e-ISBN-13: 978-3-642-80657-5
DOI: 10.1007/978-3-642-80657-5

Softcover reprint of the hardcover 1st edition 1971

TABLE OF CONTENTS

PREFACE

These notes are the result of an interrupted sequence of seminars on optimization theory with economic applications starting in 1964-1965. This is mentioned by way of explaining the uneven style that pervades them. Lately I have been using the notes for a two semester course on the subject for graduate students in economics. Except for the introductory survey, the notes are intended to provide an appetizer to more sophisticated aspects of optimization theory and economic theory.

The notes are divided into three parts. Part I collects most of the results on constrained extrema of differentiable functionals on finite and not so finite dimensional spaces. It is to be used as a reference and as a place to find credits to various authors whose ideas we report.

Part II is concerned with finite dimensional problems and is written in detail. Needless to say, my contributions are marginal. The economic examples are well known and are presented by way of illustrating the theory.

Part III is devoted to variational problems leading to a discussion of some optimal control problems. There is a vast amount of literature on these problems and I tried to limit my intrusions to explaining some of the obvious steps that are usually left out. I have borrowed heavily from Akhiezer [1], Berkovitz [7], Bliss [10] and Pars [40]. The economic applications represent some of my work and are presented in the spirit of illustration.

During the process of preparing for these notes, starting in Minnesota and ending in Kansas via Purdue, Missouri and Cairo, I have benefited a great deal from my teachers and colleagues. I am most indebted to Professors: M. Beckmann, W. Brock, A. Camacho, J. Chipman, M. El-Imam, L.S. Fan, C. Hildreth, L. Hurwicz, H. Kanemitsu, D. Katzner, M. Khairat, W.S. Loud, J. Moore, J. Quirk, S. Reiter, R. Saposnik, A. Takayama and R. Thompson. They must receive the wages of their sins of giving, wittingly or unwittingly, implicitly or explicitly, aid and comfort to the author. None of them, however, have seen the final draft of these notes. This, we hope, is grounds for clemency. I alone am responsible for all errors.

The preparation of the final draft was done at the University of Kansas, and I gratefully acknowledge the help and support that I received. Finally, it is a pleasure to express my thanks to Mrs. Alice Milberger for the magical job of transforming various bits and pieces of handwriting to a neat typescript.

Mohamed El-Hodiri
Lawrence, Kansas
July 1971

"Now you shall hear the story of Sordello,"

Browning: Sordello

I

INTRODUCTORY SURVEY

CHAPTER 1

A SURVEY OF DERIVATIVE CHARACTERIZATION OF CONSTRAINED EXTREMA

1. Introduction

In this chapter we review the theorems that characterize optimality by way of derivatives. First we formulate a very general optimization problem. Then we present characterization theorems for three types of problems: Finite dimensional, variational and problems in linear topological spaces. In each case we present theorems for equality - inequality constraints. The theorems in each case are: first order necessary conditions, first order sufficient conditions, second order necessary conditions and second order sufficient conditions. The scheme of representation is as follows: Statements of theorems are followed by remarks referring the reader to the earliest, known to us, proofs of the theorems. In some instances, slight generalizations of some theorems appear here for the first time, an indication of necessary modifications to existing proofs are presented. A case is "solved" if proofs for all the four types of characterization theorems exist. The only "unsolved" case is that of problems in linear topological spaces with inequality and with equality - inequality constraints. For this case, we present two conjectures about second order conditions that are analogous to the equality constraint case.

We now state the "general" optimization problem:

Let A,B,C,D be real linear topological spaces. Consider the functions:

$f\colon\ A \to B$

$g\colon\ A \to C$

$h\colon\ A \to D$

Let P be a partial ordering of B, let $\geq$ be defined on C - as usual - by way of a convex cone and let θ_1, θ_2, θ_3 and θ_4 be the neutral element of addition - zero elements - of A,B,C and D respectively. The problem may be stated as follows[1]:

[1] This general statement of the problem is due to Hurwicz [27].

Find $\hat{x} \in A$ such that $f(\hat{x})$ is P-maximal[2] subject to: $g(x) = \theta_3$ and $h(x) \geq \theta_4$

In case B is the real line and P is the relation "$\geq$" defined for real numbers, we have a problem of scalar optimization. In case B is finite dimensional we have a finite-vector maximization problem. In general, characterizations of solutions to finite-vector maximization problems may be derived from characterizations of solutions of scalar maximization problems. We shall restrict our presentation to scalar maximization problems. As an application of scalar maximization theorems we may "solve" a particular finite-vector maximization problem, namely for the case where P is taken to be a Pareto[3] ordering of B. For infinite-vector maximization problems, a scalarization theorem appears to be the most appropriate intermediate step to derive characterization theorems. Such a theorem was proved by Hurwicz [27]. The method, for finite-vector maximization, consists[4] of simply observing that the problem is equivalent to a finite number of scalar maximization problems. However, we shall only be concerned with scalar maximization problems in this chapter.

Finally, at the risk of blabbering the obvious, these things should be pointed out. First, the choice of differentiable functions as the class to which the results apply does not mean that we think them to be the most general class functions nor does it mean that the proofs of the theorems are the most elegant. In fact, such considerations are traded for unity and simplicity of the representation. Second, the division of problems to finite dimensional, variational, and problems in linear topological spaces. Had it been true that the problem of part 3 of this chapter is "solved", there might be a case for disregarding sections 1 and 2, for then they would be of only historical interest. But even then, one would be faced with the problem of assuming too much in order to get the more general results and obtain others as special cases. An illustration of the point is the relation between optimal con-

---

[2] We shall restrict our attention to maximization. Characterization of solutions to minimization problems follow trivially from maximization theorems.

minimizatio

[3] In that case, x is said to be Pareto superior to y if $f^i(x) \geq f^i(y)$ for $i = 1, \ldots, r$ (r is the finite dimension of B), and $\hat{x}$ is Pareto optimal (P-maximal) if there does not exist a point $y \in A$ satisfying $g(y) \geq \theta_2$, $h(y) = \theta_3$ which is Pareto superior to x.

[4] Suggested to the author by Hurwicz, see chapter 4 of these notes.

trol theory, Part 2, and Part 3, functional analysis. There exist excellent works relating the problem of maximizing a functional subject to functional constraints to optimal control problems, see e.g. [11] and [39]. We decided not to include these papers in our survey because the results, on functional analysis, were rather specialized, i.e. they apply to functionals in general but they are more suited to variational problems. One other reason for not dealing with variational problems as problems in Linear Topological Spaces is that the theory, in Part 3, is incomplete, while variational theory is.

Third, since we are only concerned with characterization, computational results and existence theory are not reported on.

2. Finite Dimensional Problems

In this part we let $A = E^n$, $B = E^1$, $C = E^m$ and $D = E^\ell$, where E^n, E^1, E^m, E^ℓ are Euclidian spaces of dimensions n, one, m and ℓ respectively. For this part we reformulate the scalar maximization problem as follows:

Problem 1. Find $\hat{x} \in E^m$ such that $f(\hat{x}) \geq f(x)$ for all x satisfying $g^\alpha(x) = 0$, $\alpha = 1, \ldots, m$, $h^\beta(x) \geqq 0$, $\beta = 1, \ldots, \ell$

We further state some properties of the constraints that will be used in discussing the theorems in this section. Some of the nomenclature, designated by " ", is Karush's [31].

D.1) Definition. Effective constraints. Let $\overset{\circ}{x}$ be a point that satisfies $h^\beta(\overset{\circ}{x}) \geq 0$, $\beta = 1, \ldots, m$. Let $\Gamma(\overset{\circ}{x})$ be the set of indices β such that $h^\beta(\overset{\circ}{x}) = 0$. The constraints with indices $\beta \in \Gamma(\overset{\circ}{x})$ will be referred to as the effective constraints at $\overset{\circ}{x}$.

D.2) Definition. "<u>Admissible Direction</u>". Let h be differentiable. We say that λ is an admissible direction if λ is a non-trivial solution to the inequalities:

$$\sum_{i=1}^{n} \overset{\circ}{h}{}^\beta_i \lambda_i \geq 0, \quad \beta \in \Gamma(\overset{\circ}{x})$$

where $\overset{\circ}{h}{}^\beta_i = \frac{\partial h^\beta}{\partial x_i} \mid x = \overset{\circ}{x}$.

D.3) Definition. "A curve issuing from $\overset{\circ}{x}$ in the direction λ." By that we mean, an n-vector valued function, $\xi(t)$, of a real variable t such that $\xi(0) = \overset{\circ}{x}$ and

$$\xi'(0) = \frac{d}{dt}\,\xi(t)\Big|_{t=0} = \lambda.$$

D.4) Definition. "An admissible arc issuing from $\overset{\circ}{x}$ in the direction λ" is an arc issuing from $\overset{\circ}{x}$ in the direction λ such that $h(\xi(t)) \geq 0$.

D.5) Definition. "Property Q" for inequality constraints is satisfied at $\overset{\circ}{x}$. Iff: For each admissible direction λ, there exists an admissible arc issuing from $\overset{\circ}{x}$.

D.6) Definition. The rank condition for inequality constraints. We say that the rank condition for inequality constraints is satisfied at $\overset{\circ}{x}$ iff: 1) the constraints are differentiable and 2) the rank of the matrix $[\overset{\circ}{h}{}^{\beta}_{i}]$, $\beta \in \Gamma(\overset{\circ}{x})$, equals the number of effective constraints, i.e., is maximal.

D.7) Definition. The rank condition for equality-inequality constraints is satisfied at $\overset{\circ}{x}$ iff: 1) the functions g and h are differentiable and 2) the rank of the matrix $\begin{bmatrix} \overset{\circ}{g}{}^{\alpha}_{i} \\ \overset{\circ}{h}{}^{\beta}_{i} \end{bmatrix}_{,\beta \in \Gamma(\overset{\circ}{x})}$ equals ℓ + the number of effective constraints at $\overset{\circ}{x}$, where $\overset{\circ}{g}_{i} = \dfrac{\partial g^{\alpha}}{\partial x_i}\Big|_{x = \overset{\circ}{x}}$.

D.8) Definition. The rank condition for equality constraints is satisfied at $\overset{\circ}{x}$ if the rank of the matrix $[\overset{\circ}{g}{}^{\alpha}_{i}]$ is α.

Finally we define local and global constraint maxima:

D.9) Definition. $\hat{x}$ is said to be a local solution of problem 1, iff: There exists a neighborhood of $\hat{x}$, $N(\hat{x})$ such that $f(\hat{x}) \geq f(x)$ for all $x \in N(x)$ satisfying $g(x) = 0$ and $h(x) \geq 0$.

D.10) Definition. $\hat{x}$ is said to be a global solution to problem 1, iff: $f(\hat{x}) \geq f(x)$ for all x satisfying $g(x) = 0$ and $h(x) \geq 0$.

2.1 First order necessary conditions.

Theorem 1. If f, g and h are continuously differentiable and if $\hat{x}$ is a global solution to problem 1, then there exists a vector $(\lambda_o, \nu, \mu) = (\lambda_o, \nu^1, \ldots, \nu^m, \mu^1, \ldots, \mu^{\ell}) \neq 0$ such that $\lambda_o \geq 0$, and:

1) $\mu^{\beta} \geq \mu^{\beta} h^{\beta}(\hat{x}) = 0$

2) $F^{o}_{x} = 0$, where $F^{o} = \lambda_o f + \nu \cdot g + \mu \cdot h$, and F_x is the vector of partial derivatives of F with respect to the components of x evaluated at $x = \hat{x}$.

For the case of equality constraints, i.e. the set $\{x|h(x) \geq 0\} = E^n$, Theorem 1 was proved by Caratheodory[5] [14] and Bliss[6] [9]. For the case of inequality constraints, i.e. the set $\{x|g(x) = 0\} = E^n$, Theorem 1--except for the non-negativity of μ in condition 1 of the conclusion--was proved by Karush[7] [31]. The non-negativity of μ may be proved by using a separation theorem[8]. This was, essentially, the crux of Fritz John's [28] proof who has treated a more general problem of an infinite number of inequality constraints. For the equality-inequality constraints the proof involves writing equality constraints as inequality constraints ($g(x) \geq 0$ and $-g(x) \geq 0$) and applying Karush's of Fritz John's theorem.

Theorem 2. If, in addition to the assumption of Theorem 1 we have either:[9]

a) Property Q for inequality constraints, D.4, and the rank condition for equality constraints, D.7, are satisfied at $\hat{x}$.

a') The rank condition for equality-inequality constraints is satisfied at $\hat{x}$.

Then the conclusions of Theorem 1 follow with $\lambda_o \neq 0$.

For equality constraints the theorem was proved as a corollary of Theorem 1 by Bliss[10] [9]. For the case of inequality constraints the theorem was proved by Karush[11] [31]. The proof, for equality-inequality constraints, may be accomplished by converting inequality constraints to equality constraints and obtaining the theorem as a corollary to Theorem 1. This was presented by Pennisi[13] [41]. A direct

---

[5] Theorem 2, section 187 part II.

[6] Theorem 1.1.

[7] Theorem 3.1. Karush proves the non-negativity of μ by using the second order necessary condition, thus necessitating the assumption that the functions (maximal and constraints) have second order continuous partial derivatives. This, obviously, is a dispensible assumption.

[8] Separating the linear sets:

$\hat{f}_i\xi_i < 0$ and $\overset{o\beta}{h}_i\xi_i \geq 0$. See Dubovskii & Milyutin [16] for an extensive discussion.

[9] The conditions that follow are alternative forms of the constraint qualification, see Arrow-Hurwicz, Uzawa [3] for other forms and for relations among various forms of the constraint qualification.

[10] In remarks following Theorem 1.1.

[11] Theorem 3.2.

[12] Following Karush [31], by writing $h^\beta(x) \geq 0$ as $h^\beta(x) - (z^\beta)^2 = 0$ and solving the problem in the space of $n + m$ - vectors $(x, \bar{z})$.

[13] Theorem 3.1.

proof was presented by Hestenes[14] [26].

2.2 First order sufficient conditions.

Theorem 3. If: 1) f, g and h are differentiable. 2) The conclusions of Theorem 1 hold with $\lambda_o > 0$ at a point x with g(x) = 0, h(x) 0. 3) Either (3.a) F^o, of Theorem 1, is concave or (3.b) g is linear and f and h are concave. Then x is a global solution to problem 1.

The theorem follows from the fact that a concave function lies below its tangent plane. The implications of this fact were utilized by Kuhn-Tucker [32] in the proof of their equivalence theorem. The present theorem may be proved by applying Lemma 3 of Kuhn and Tucker [32] to F^o.

2.3 Second order necessary conditions.

Theorem 4. If: 1) f, g and h have second order continuous partial derivatives. 2) $\hat{x}$ is a solution to problem 1. 3) The rank condition for equality-inequality constraints, D.6, is satisfied.

Then there exist multipliers $(\nu, \mu) = (\nu^1, \ldots, \nu^m; \mu^1, \ldots, \mu^\ell)$ such that:

The conclusions of Theorem 2 hold and $\Sigma_{i,j} \hat{F}_{ij} \eta_i \eta_j \leq 0$ for all $\eta = (\eta_1, \ldots, \eta_n) \neq 0$ with $\Sigma_i \hat{g}_i^\alpha \eta_i = 0$, $\Sigma_i \hat{h}_i^\beta \eta_i = 0$, $\alpha = 1, \ldots, m$, $\beta \in \Gamma(\hat{x})$ where, $F = f + \Sigma_\alpha \nu^\alpha g^\alpha + \Sigma_\beta \mu^\beta h^\beta$ and $\hat{F}_{ij} = \left.\frac{\partial^2 F}{\partial x_i \partial x_j}\right|_{x = \hat{x}}$.

For the equality constraints, the theorem was proved by Caratheodory[15] [14] and Bliss[16] [9]. For inequality constraints the theorem was proved by Karush[17] [31]. For the equality-inequality case, the theorem was proved by Pennisi[18][41], under the, dispensible[19], stipulation that the number of non-zero multipliers attached to effective inequality constraints is at most one. A direct proof was presented by Arrow and Hurwicz, Theorem 3 in: K. Arrow and L. Hurwicz [2].

---

[14] Theorem 10.1, Chapter 1.

[15] Theorem 3, section 212, part II.

[16] Theorem 1.2.

[17] Theorem 5.1.

[18] Corollary to Theorem 3.2.

[19] See Chapter 3, section 3 of these notes, where Pennisi's theorem is proved by directly applying the second order necessary conditions for equality constraints using Karush's device (footnote 12).

2.4 Second order sufficient conditions.

Theorem 5. If: 1) f, g, and h have continuous second order derivatives. 2) The conclusions of Theorem 2 hold at $\hat{x}$ with $g(\hat{x}) = 0$, $h(\hat{x}) \geq 0$. 3) The conclusion of Theorem 4 holds with strict inequality for $\eta \neq 0$. Then $\hat{x}$ is a local solution to problem 1.

For equality constraints the theorem was proved by Bliss[20] [9]. Caratheodory[21] [14] assumes, needlessly, that the rank condition holds. For inequality constraints the theorem was proved by Karush[22] [31]. A very closely related theorem was proved by Pennisi[23] [41], where the restrictions on η are augmented by requiring that $\Sigma_i \hat{h}_i^\beta \eta_i \geq 0$ for indices β with $\mu^\beta > 0$. Theorem 5 may be proved by applying the sufficiency theorem for equality constraints by using Karush's device[24] [31]. A direct proof was provided by Hestenes[25] [26].

2.5 Second order conditions in terms of determinants.

Let the matrix $H = \begin{bmatrix} H_1 & H_2 \\ H_3 & H_4 \end{bmatrix}$ be defined as follows: $H_1 = [\hat{F}_{ij}]$, where $\hat{F}_{ij}$ is as defined in Theorem 4, $H_2 = H_3' = \begin{bmatrix} \hat{g}_i^\alpha \\ \hat{h}_i^\beta \end{bmatrix}'_{\beta \in \Gamma(\hat{x})}$, with primes denoting transposes and using the notation of definition D.7 and where H_4 is a square matrix of zeros of order $b = \ell$ + the number of effective constraints at $\hat{x}$. Let $H^{(k)}$ denote bordered principle minors of order k of H defined as follows:

1) Eliminate all rows and columns of H_1 except the first k, 2) Eliminate all rows of H_2 except the first k , 3) Eliminate all columns of H_3 except for the first k, 4) Keep H_4 as it is.

Corollary to Theorem 4. Under the assumptions of Theorem 4, the bordered principle minors of H of orders k, ranging from $b = \ell$ + the number of effective constraints at $\hat{x}$ to n, alternate in signs. Some or all may be zero. The first bordered principle minor, of order b, has the sign of $(-1)^{b+1}$.

---

[20] Theorem 1.3.

[21] Theorem 4, section 213, part II.

[22] Theorem 6.1.

[23] Theorem 3.3.

[24] See chapter 3 of these notes.

[25] Theorem 10.3, chapter 1.

Corollary to Theorem 5. <u>Conditions 3 of Theorem 5 may be replaced by: None of the bordered principle minors of H of order $b < k < n$ is zero and they alternate in sign with the first, of order b, having the sign of $(-1)^{b+1}$.</u>

Both corollaries follow from Caratheodory's results, sections 209, 213-217 of [14], on necessary and sufficient conditions for a quadratic form to be semi-definite (definite) under linear constraints.

3. A Variational Problem

In view of the vast literature on the calculus of variations and the wide accessability of such literature, this part will be very brief. Many interesting topics will be left out, e.g. problems with retarded arguments[26]. For a survey of variational problems with equality constraints, the reader is referred to Bliss [8], which we shall take as a point of departure for this part. We formulate the Bolza-Hestenes[27] problem in the calculus of variations as follows: Let T be a subset of the non-negative half of the real line. Consider the class of piecewise smooth functions x(t) defined on t and having values in E^n together with their derivatives $\dot{x}$. The problem is:

<u>Problem 2.</u> Find $\hat{t}_o$, $\hat{t}_1$, $\hat{\dot{x}}(t)$, $\hat{x}(\hat{t})$, $\hat{x}(\hat{t}_o)$, $\hat{x}(\hat{t}_1) = \hat{z}$ that maximizes $J^o[z] =$

$$\int_{t_o}^{t_1} f^o(t, x, \dot{x})dt + g^o(t_o, t_1, x(t_o), x(t_1)) \text{ subject to}$$

(1) $J^{\alpha}[z] = \int_{t_o}^{t_1} f^{\alpha}(t, x, \dot{x})dt + g^{\alpha}(t_o, t_1, x(t_o), x(t_1)) = 0, \ \alpha = 1, \ldots, m_1,$

(2) $J^{\bar{\alpha}}[z] = \int_{t_o}^{t_1} f^{\bar{\alpha}}(t, x, \dot{x})dt + g^{\bar{\alpha}}(t_o, t_1, x(t_o), x(t_1)) \geq 0, \ \bar{\alpha} = m_1 + 1, \ldots, m,$

(3) $\phi^{\beta}(t, x, \dot{x}) = 0, \ \beta = 1, \ldots, \ell_1,$

(4) $\phi^{\bar{\beta}}(t, x, \dot{x}) \geq 0, \ \bar{\beta} = \ell_1 + 1, \ldots, \ell,$

(5) $\psi^{\gamma}(t_o, t_1, x(t_o), x(t_1)) = 0, \ \gamma = 1, \ldots, S_1,$

(6) $\psi^{\bar{\gamma}}(t_o, t_1, x(t_o), x(t_1)) \geq 0, \ \bar{\gamma} = S_1 + 1, \ldots, S$, where $t_o, t_1 \in T$ with $t_o < t_1$.

As was noted, by Berkovitz [7] and Hestenes [25], the above problem is equivalent to the problem of optimal control[28]. The results of optimal control theory are

---

[26] See El'sgol's [18], Halanay [24], Ewing [19], and chapter 7 of these notes.

[27] see Hestenes [25].

[28] As stated in these two papers.

derivable from the results that we shall present.[29]

3.1 First order necessary conditions.

Theorem 6. If: 1) The functions f, g, ϕ and ψ are continuously differentiable, as functions of real variables. 2) $\hat{z}$ is a solution to problem 2. 3) The matrix $\begin{bmatrix} \hat{\phi}^{\beta}_{\dot{i}} \\ \hat{\phi}^{\bar{\bar{\beta}}}_{\dot{x}_i} \end{bmatrix}$ has full rank at each point $t \in [\hat{t}_o, \hat{t}_1]$, where $\hat{\phi}^{\beta}_{\dot{x}_i} = \left.\frac{\partial \phi^{\beta}}{\partial \dot{x}_i}\right|_{z = \hat{z}}$, $\hat{\phi}^{\bar{\bar{\beta}}}_{\dot{x}_i} = \left.\frac{\partial \phi^{\bar{\beta}}_{\dot{i}}}{\partial \dot{x}_i}\right|_{z = \hat{z}}$ and where the index $\bar{\bar{\beta}}$ designates indices $\bar{\beta}$ with $\phi^{\bar{\beta}}(\hat{z}) = 0$, $i = 1, \ldots, n$.

Then there exists a constant vector $(\lambda_o, q, \omega) = (\lambda_o; q^1, \ldots, q^m; \omega^1, \ldots, \omega^s)$ and a vector function $p = (p^1, \ldots, p^{\ell})$ defined on $[\hat{t}_o, \hat{t}_1]$ such that:

1) There does not exist $t \in [\hat{t}_o, \hat{t}_1]$ with $[\lambda_o, q, \omega, p] = 0$, also $(\lambda_o, q, \omega) \neq 0$,

2) $\lambda_o \geq 0$, $q^{\bar{\alpha}} \geq 0$, $q^{\bar{\alpha}} J^{\bar{\alpha}}[\hat{z}] = 0$,

3) P(t) is piecewise continuous, continuous at points of continuity of $\dot{x}$, $p^{\bar{\beta}}(t) \geq 0$, $p^{\bar{\beta}}(t)\ \phi^{\alpha}[\hat{z}] = 0$,

4) $\omega^{\gamma} \geq 0$, $\omega^{\gamma}\psi^{\gamma}[\hat{z}] = 0$

5) $\frac{d}{dt} \hat{F}_{\dot{x}_i} = \hat{F}_{x_i}$, $i = 1, \ldots, n$, where $F = \lambda_o f^o + \sum_{\alpha} q^{\alpha} f^{\alpha} + \sum_{\beta} \mu^{\beta}\phi^{\beta}$,

6) $\frac{d}{dt}(\hat{F} - \sum_i \hat{\dot{x}}_i \hat{F}_{\dot{x}_i}) = \frac{\partial}{\partial t}\hat{F}$,

7) $dG + [(\hat{F} - \sum_i \hat{\dot{x}}_i \hat{F}_{\dot{x}_i})dt_{\gamma} + \sum_i \hat{F}_{\dot{x}_i}\, dx_i(t_{\gamma})]_{\gamma=o}^{\gamma=1} = 0$ is an identity in $dx_i(t_{\gamma})$, dt_{γ}, $\gamma = 0, 1$; $i = 1, \ldots, n$; where $G = \lambda_o g^o + \sum_{\alpha} q^{\alpha} g^{\alpha} + \sum_{\gamma} \omega^{\gamma}\psi^{\gamma}$,

8) $E(t, \hat{x}, \hat{\dot{x}}, \dot{x}) = F(t, \hat{x}, \hat{\dot{x}}) - F(t, \hat{x}, \dot{x}) - \sum_i (\hat{\dot{x}}_i - \dot{x}_i)\hat{F}_{\dot{x}_i} \geq \sum_{\bar{\beta}} \mu^{\bar{\beta}}\phi^{\bar{\beta}}$ whenever $(t, \hat{x}, \dot{x})$ satisfy the constraints (1) - (6).

For fixed end points (i.e. t_o, t_1, $x(t_o)$ end $x(t_1)$ are constants), Valentine [48] proved that: conclusion 5) and 8)[30-31] are necessary with conclusion 3) holding for

---

[29] See Berkovitz [7] and Hestenes [25] and chapter 7 of these notes for the necesary transformations.

[30] First Necessary Condition I, Page 412.

[31] Second Necessary Condition II, Page 414.

p except for the non-negativity of $p^{\bar{\beta}}$. The non-negativity of $p^{\bar{\beta}}$ was proved by applying a Clebsch[32] type second order necessary condition[33]. Valentine's method consisted of converting differential inequality constraints to equations by subtracting from each, the square of a derivative of an added variable. Then he derived his results as applications of characterization theorems for the problem of Balza with equality constraints[34]. For the general problem we may convert the inequality constraints (2) and (6) to equality constraints by introducing three new sets of variables $y_1^{\bar{\alpha}}(t)$, $y_2^{\bar{\beta}}(t)$ and $y_3^{\bar{\gamma}}(t)$ as follows : $\dot{y}_1^{\bar{\alpha}} = 0$, $y_1^{\bar{\alpha}}(t_1)$ free, $\dot{y}_2^{\bar{\gamma}}(t) = 0$, $y_2^{\bar{\beta}}(t\ )$ free and $\dot{y}_3^{\bar{\gamma}} = 0$, $y_3^{\bar{\gamma}}(t_1)$ free, and consider the equivalent problem of maximizing J^o subject to constraints (1), (3), (5) and: (2') $\bar{J}^{\bar{\alpha}} = J^{\bar{\alpha}} - (\bar{y}_1^{\bar{\alpha}}(t_1))^2 = 0$, (4') $\bar{\phi}^{\bar{\beta}} = \bar{\phi}^{\bar{\beta}} - (\dot{y}_2^{\beta})^2 = 0$ and (6') $\bar{\psi}^{\bar{\gamma}} = \psi^{\bar{\alpha}} - (y_3^{\bar{\alpha}}(t_1))^2 = 0$. We then get, in addition to Valentine's conditions, condition 7) of Theorem 6. Noting that condition 6) may be obtained from 5) and 7) we would have all of the conditions of the theorem. A direct elegant proof of Theorem 6 was provided by Hestenes [25].

We now discuss a condition that guarantees that λ_o in Theorem 6 is non-zero. The case where the multipliers are unique, choosing $\lambda_o = 1$, is what is known in the literature as the normal case. Although conditions for normality are very hard to verify in applications, we shall present one of these conditions in this section[35].

D.11) Definition. Normality. $\hat{z}$ is said to be normal iff conclusions 1) - 5) and 7) of Theorem 6 hold with $\lambda_o = 1$ and the multipliers q, p, ω are unique.

D.12) Definition. Admissible variations. Consider a point $\bar{z}$ and a vector valued functions $\xi^\sigma(t) = (\xi_1^\sigma(t), \ldots, \xi_n^\sigma(t))$; $\sigma = 1, \ldots, \bar{m} + \bar{s}$, where $\bar{m} = m_1$ + the number of constraints $J^{\bar{\alpha}}$ that hold as equations at $\bar{z}$, effective at the end points of $\bar{z}$, and where $\bar{s} = s$ + the number of constraints $\psi^{\bar{\gamma}}$ that are effective at $\bar{z}$. $\xi(t) = (\xi^1(t), \ldots, \xi^{\bar{m}+\bar{s}}(t))$ is said to be admissible variations iff:

1) $\xi_i(t)$ are differentiable on $[\bar{t}_o, \bar{t}_1]$,

---

[32] See section 3.3 in this chapter.

[33] Corollary 3.4.

[34] These theorems may be found in Bliss's paper [9].

[35] See Berkovitz [7], section VIII, theorem 3, for alternative sufficient condition for normality.

2.1) $\phi^{\beta}(\xi^{\sigma}) = \sum_{i=1}^{m} \phi^{\beta}_{x_i} \xi^{\sigma}_i(t) + \sum_{i=1}^{m} \phi^{\beta}_{\dot{x}_i} \dot{\xi}^{\sigma}_i(t) = 0, \beta = 1, \ldots, m_1,$

2.2) $\phi^{\bar{\bar{\beta}}}(\xi^{\alpha}) = \sum_i \phi^{\bar{\beta}}_{x_i} \xi^{\sigma}_i + \sum_i \phi^{\bar{\beta}}_{x_i} \dot{\xi}^{\sigma}_i = 0$ for $\bar{\bar{\beta}}$ with $\phi^{\beta}(\bar{z}) = 0$, where $\phi^{\beta}_{\dot{x}_i} = \left.\frac{\partial\phi^{\beta}}{\partial x_i}\right|_{z = \bar{z}}$ and where $\phi_{\dot{x}_i}$, $\phi^{\bar{\bar{\beta}}}_{x_i}$ and $\phi^{\bar{\bar{\beta}}}_{\dot{x}_i}$ are defined similarly.

D.13) Definition. <u>The rank condition</u>. The first rank condition is said to be satisfied at $\bar{z}$ if there exists a set of admissible variations ξ^{σ}, and arbitrary constants τ^{σ}_o, τ^{σ}_1, $\sigma = 1, \ldots, \bar{m} + \bar{s}$, such that the matrix $\begin{bmatrix}\bar{L}_1\\ \bar{L}_2\\ \bar{L}_3\\ \bar{L}_4\end{bmatrix}$ has rank $\bar{m} + \bar{s}$, where

$$\bar{L}_1 = [\bar{L}^{\alpha\sigma}_1] = [(g^{\alpha}_{t_o} + \sum_i g^{\alpha}_{x_{i(t_o)}} \dot{\bar{x}}_i(t_o) - f^{\alpha}(\bar{t}_o, x(t_o), x(t_o)))\tau^{\sigma}_o + \sum_i \bar{g}^{\alpha}_{x_i}(t_o)\xi^{\sigma}_i(t_o) +$$

$$+ (\bar{g}^{\alpha}_{t_1} + \sum_i \bar{g}^{\alpha}_{x_i}(t_1)\dot{x}_i(t_1) + f^{\alpha}(\bar{t}_1, \bar{x}(\bar{t}_1), \dot{\bar{x}}(\bar{t}_1)))\tau^{\sigma}_1 + \sum_i \bar{g}^{\alpha}_{x_i}(t_1)\xi^{\sigma}_i(t_1) + \int_{t_o}^{t_1}$$

$$(\sum_i \bar{f}^{\alpha}_{x_i} \xi^{\sigma}_i(t) + f^{\alpha}_{x_i} \dot{\xi}^{\sigma}_i(t))\, dt],\ \alpha = 1, \ldots, m_1,\ \sigma = 1, \ldots, \bar{m} + \bar{s},$$

$$\bar{L}_2 = [\bar{L}^{\bar{\bar{\alpha}}\sigma}_2] = [\bar{g}^{\bar{\bar{\alpha}}}_{t_o} + \sum_i \bar{g}^{\bar{\bar{\alpha}}}_{\dot{x}_i}(t_o)\dot{\bar{x}}(t_o) - \bar{f}^{\bar{\bar{\alpha}}}(\bar{t}_o, \bar{x}(\bar{t}_o), \dot{\bar{x}}(\bar{t})))\tau^{\sigma}_o + \sum_i \bar{g}^{\bar{\bar{\alpha}}}_{x_i}(t_o)\xi^{\sigma}_i(t_o) +$$

$$+ (\bar{g}^{\bar{\bar{\alpha}}}_{t_1} + \sum_i \bar{g}^{\bar{\bar{\alpha}}}_{x_i}(t_1)\dot{\bar{x}}(t_1) + f^{\bar{\bar{\alpha}}}(\bar{t}_1, \bar{x}(\bar{t}_1), \dot{\bar{x}}(\bar{t}_1)))\tau^{\sigma}_1 + \sum_i \bar{g}^{\bar{\bar{\alpha}}}_{x_i}(t_1)\xi^{\sigma}(t_1) + \int_{\bar{t}_o}^{\bar{t}_1}$$

$$(\sum_i \bar{f}^{\bar{\alpha}}_{x_i} \xi^{\sigma}_i(t) + \bar{f}^{\alpha}_{x_i} \xi^{\sigma}_i(t))\, dt],$$ $\bar{\bar{\alpha}}$ denotes indices of effective constraints $J[\bar{z}]$ at $\bar{z}$, $\bar{\sigma} = 1, \ldots, m + s$,

$$L_3 = [\bar{L}^{\gamma\sigma}_3] = [\bar{\psi}^{\gamma}_{t_o} + \sum_i \bar{\psi}^{\gamma}_{x_i}(t_o)\dot{\bar{x}}_i(t_o))\tau^{\sigma}_o + \sum_i \bar{\psi}^{\gamma}_{x_i}(t_o)\xi^{\sigma}_i(t_o) + (\bar{\psi}^{\gamma}_{t_1} + \sum_i \psi^{\gamma}_{x_i} \dot{\bar{x}}(t_1))\tau^{\sigma}_1 +$$

$$+ \bar{\psi}^{\gamma}_{x_i}(t_1)\xi^{\sigma}_i(t_1)],\ \gamma = 1, \ldots, s_1,\ \sigma = 1, \ldots, \bar{m} + \bar{s},$$

$$\bar{L}_4 = [\bar{L}^{\bar{\bar{\gamma}}\sigma}_4] = [\bar{\psi}^{\bar{\bar{\gamma}}}_{t_o} + \sum_i \bar{\psi}^{\bar{\bar{\gamma}}}_{x_i(t_1)} \dot{\bar{x}}(t_o))\, \tau^{\sigma}_o + \sum_i \bar{\psi}^{\bar{\bar{\gamma}}}_{x_i(t_o)} \xi^{\sigma}_i(t_o) +$$

$$(\bar{\psi}^{\bar{\bar{\gamma}}}_{t_1} + \sum_i \bar{\psi}^{\bar{\bar{\gamma}}}_{x_i(t_1)} \dot{\bar{x}}(t_1))\, \tau^{\sigma}_1 + \sum_i \bar{\psi}^{\bar{\bar{\gamma}}}_{x_i(t_1)} \xi^{\sigma}_i(t_1)],$$ $\bar{\bar{\gamma}}$ denotes indices of $\psi^{\bar{\gamma}}$ that are effective at $\bar{z}$, $\sigma = 1, \ldots, \bar{m} + \bar{s}$, and where subscripts denote partial

derivatives with respect to indicated variables and " — " above an expression indicates that it is evaluated at $\bar{z}$.

Theorem 7. If f, g, ϕ and ψ are differentiable then the rank condition, definition D.12., at $\bar{z}$ is necessary and sufficient for the normality of $\bar{z}$.

In the absence of inequality constraints, the theorem was proved by Bliss[36] [8]. Reformulating the problem with added inequality constraints, as indicated above, the theorem is obtained as a straight forward application of Bliss's theorem.

3.2 First order sufficient conditions.

Theorem 8. If 1) f, g, ϕ and ψ are differentiable, and concave. 2) Conclusions 1) - 5) and 7) of Theorem 6 are satisfied with: 2.i) $\lambda_o > 0$, 2.ii) $q^\gamma \geq 0$, $\omega^\gamma \geq 0$, $\gamma = 1, \ldots, m_1$, $\gamma = 1, \ldots, s_1$, at a point $\hat{z}$ that satisfies the constraints (1) - (6). Then $\hat{z}$ is a global solution to problem 2.

The theorem was proved by Mangasarian [38] for the canonical, optimal control, problem with fixed t_o and t_1. Theorem 8 may be proved by repeating the steps of Mangasarian's proof for problem 2. As Mangasarian notes, in the absence of equality constraints condition 2.ii) of Theorem 8 is not needed, nor is it required if the equality constraints are linear.

3.3 Second order necessary conditions.

Theorem 9. (Jacobi-Myer-Bliss) If: 1) f, g, h and ψ have continuous second order derivatives, 2) $\hat{z}$ is a solution to problem 2 and 3) $\hat{z}$ is normal. Then there exist multipliers as in Theorem 6 with $\lambda_o = 1$ such that: $Q(\hat{z},\tau,\xi) = d^2\, G(\hat{z};\tau,\xi) +$

$$[(\hat{F}_t - \sum_i \hat{F}_{x_i} \dot{x}_i)\, \tau_\gamma^2 + 2 \sum_i \hat{F}_{x_i}\, \xi_i(t\gamma)\, \tau_\gamma)]\Big|_{\gamma=0}^{\gamma=1} + \int_{\hat{t}_o}^{\hat{t}_1} (\sum_{i,j} \bar{\hat{F}}_{x_i x_j}\, \xi_i \xi_j +$$

$$2 \sum_{i,j} \hat{F}_{x_i \dot{x}_j}\, \xi_i \dot{\xi}_j + \sum_{i,j} \hat{F}_{\dot{x}_i \dot{x}_j}\, \dot{\xi}_i \dot{\xi}_j)\, dt \leq ,$$ for $(\tau, \xi) = (\tau_o, \tau_1, \xi_1(t), \ldots, \xi_n(t)) \neq 0$

satisfying

(i) $\hat{\phi}^\beta(\xi) = 0$, $\hat{\phi}^\beta(\xi) = 0$

(ii) $\hat{L}_1^\gamma(\tau, \xi) = 0$, $\overset{=}{\hat{L}}{}_2^\gamma(\tau, \xi) = 0$, $\hat{L}_3^\gamma(\tau, \xi) = 0$, $\overset{=}{\hat{L}}{}_4^\gamma(\tau, \xi) = 0$,

---

36 Section 9, Page 693.

with the terms in (i) and (ii) defined as in (2) of D.12 and as in D.13 with (τ, ξ) as a matrix with one row $(\tau^\sigma, \xi^\sigma) = (\tau, \xi)$, and where: $d^2G(\hat{Z}; \tau, \xi)$ is the second differential of G at $\hat{Z}$ with (τ, ξ) as increments, a single subscript denotes a first derivative and a double subscript denotes second (mixed partial) derivatives with " ^ " signifying evaluation at $\hat{Z}$.

For equality constraints the theorem was proved by Bliss[37] [8] and [10][38]. With added inequality constraint the theorem may be proved by applying Bliss's theorem to our problem, after converting it to a problem with equality constraints as we indicated in the discussion of Theorem 6; see chapter 6 of these notes.

Theorem 10. (The Clebsch condition). If the hypotheses of Theorem 9 are satisfied then $\sum_{i,j} \hat{F}_{\dot{x}_i\dot{x}_j} \Pi_i\Pi_j \leq 0$ for $\Pi = (\Pi_1, \ldots, \Pi_n) \neq 0$ satisfying $\sum_{i=1}^{n} \hat{\phi}^\beta_{\dot{x}_i} \Pi_i = 0$, $\beta = 1, \ldots, \ell_1$, $\sum_i \hat{\phi}^{\bar{\bar{\beta}}}_{\dot{x}_i} \Pi_i = 0$, $\bar{\bar{\beta}}$ are indices of inequality constraints that are effective at $\hat{Z}$, $\sum_i \hat{f}^\alpha_{\dot{x}_i} \Pi_i = 0$, $\alpha = 1, \ldots, m_1$, $\sum_i \hat{f}^{\bar{\alpha}}_{\dot{x}_i} \Pi_i = 0$, $\bar{\alpha}$ indicates constraints $J^{\bar{\alpha}}$ effective at $\hat{Z}$.

Theorem 10 was proved by Valentine[39] [48] for fixed end points. Valentine's method of proof can, easily, be applied to prove our theorem; see chapter 6 of these notes.

3.4 Second order sufficient conditions.

For the purposes of this section, we have to define a weak local solution of problem 2.

D.13. Definition. Weak Local Solution. We say that $\hat{Z}$ is a weak local solution if $J^0[\hat{Z}] \geq J^0[Z]$ for all Z satisfying the constraints (1) - (6) with $||(\hat{x}-x, \dot{\hat{x}}-\dot{x})|| < \varepsilon$ for some $\varepsilon > 0$, where $||\cdot||$ denotes the Euclidian norm.

Theorem 11. (Pennisi) If 1) f, g, ϕ and ψ have continuous second order Partial derivatives. 2) Conclusions 1) - 5) and 7) of Theorem 6 are satisfied at a point Z that satisfies the constraints (1) - (6). 3) The matrix $\begin{bmatrix} \hat{\phi}^\beta_{\dot{x}_i} \\ \hat{\phi}^{\bar{\bar{\beta}}}_{\dot{x}_i} \end{bmatrix}$, of Theorem 6, has full rank. 4) The form $Q(\hat{Z}; \tau, \xi)$ of Theorem 9 is negative definite under constraints (i)

---

[37] Sections 24 - 26.

[38] Theorem 80, Page 228, the statement and proof here are more complete than they are in [8].

[39] Corollary 3:4 Page 9.

and (ii) (in the statement of Theorem 9) for $(\tau, \xi) \neq 0$. Then there exists an $\varepsilon > 0$ such that $\hat{z}$ is a weak local solution of problem 2, in the sense of definition D.13.

Assuming that $\lambda_o = 1$ and that at most one inequality constraint is effective at $\hat{z}$, Valentine[40] proved an analogous theorem in [48]. Pennisi[41] [41] proved a sufficiency theorem without the assumptions of Valentine. Pennisi's theorem is stronger than Theorem 11 in the sense that Q is negative on the subset of variations (τ, η) of Theorem 11 which in addition (to (i) and (ii)) of Theorem 9) satisfy some inequalities for those constraints that are effective but have zero multipliers. One way to prove Theorem 11 is to convert the problem into one with equality constraints and apply Pennisi's theorem. Another way is to note that Theorem 11 is corollary of Pennisi's theorem, after modifying the latter to take care of the additional constraints. Conditions for the non-negativity of the quadratic forms in sections 3.3 and in this section, analogous to those of section 2.5, remain to be worked out.

4. A Problem In Linear Topological Spaces.

Let A, C, D be real Banach spaces and let R denote the real line. Consider f: $A \to R$, g: $A \to C$ and h: $A \to D$. The problem we study here is:

Problem 3. Maximize $f(x)$ subject to $g(x) = \theta_3$ and $h(x) \geq \theta_4$ where the inequality θ_3 and θ_4 are as defined in the introduction.

Dealing with differentiable functions we note that there are numerous equivalent[42] ways of defining derivatives in linear spaces. We shall use Frechet's definition and mean "Frechet differentiable" when we say that a function is differentiable.[43]

4.1 First order necessary conditions.

Theorem 12. If f, g and h are differentiable and if $\hat{x}$ is a solution to problem 3 then there exists a constant $\lambda_o \geq 0$ and linear functionals ℓ^1: $C \to R$, ℓ^2: $D \to R$ such that 1) $\ell^2 \geq 0$, $[\ell^2, h(\hat{x})] = 0$, where $[\ell^2, h]$ denotes the value of the functional ℓ^2 at

---

[40] Theorem 10.2, section 10.

[41] Theorem 2.1.

[42] See Averbukh and Smolyanov [4] and [5].

[43] See Vainberg [47] and Liusternik and Sobolev [34] for an exposition of calculus in linear spaces.

$h(\hat{x})$. 2) For any $y^1 \in C$, $y^2 \in D$, the triple $(\lambda_o, [\ell^1, y^1], [\ell^2, y^2] \neq 0$. 3) $F'(\hat{x}) = 0$, where $F = \lambda_o f + [\ell^1, g] + [\ell^2, h]$ and $F'(\hat{x}) = dF(\hat{x}, \xi)$ with ξ as the "increment" in the definition of the differential.

The theorem follows directly from Theorem 2.1 in Duboviskii and Milyutin[44] [16].

Conditions that guarantee that $\lambda_o > 0$ and that the functionals ℓ^1 and ℓ^2 are unique are referred to in the literature, pertaining to problem 3, as regularity conditions and as constraint qualifications. We shall now list these conditions and present some sufficient conditions for them to hold.

4.1.1 Regularity conditions.

R.1 (Gapushkin [21]): For equality constraints, $\bar{x}$ is said to (R.1)[45] regular if for every $\xi \in A$ with $g'(\bar{x}, \xi) = \theta_3$, $\xi \neq \theta_1$ we have : There exists a function of a real variable t, V: $[0, 1] \to A$ such that $V(0) = \bar{x}$, $g(V(t)) = \theta_3$ for $t \in [0, 1]$, $V'(t, \tau)$ exists for $t \in [0, 1]$ and $V'(0, \tau) = \xi$.

R.2 (Hurwicz [27]): For the inequality constraint, $\bar{x}$ is (R.2) regular iff: For any $\xi \in A$ with $\xi \neq \theta_1$ such that $x = \bar{x} + \xi$ implies $h'(\bar{x}, \xi) + h(x) \geq \theta_4$, we have: There exists a function of a real variable t, V: $[0, 1] \to A$ such that:

(i) $V'(t, \tau)$ exists for $t \in [0, 1]$

(ii) $\bar{x} = V(0)$

(iii) $h(V(t)) \geq \theta_4$, $t \in [0, 1]$

(iv) $V'(0, \tau) = \xi$, $\tau > 0$.

R.3 (Gapushkin [27]): $\bar{x}$ is said to be (R.3) regular iff: For any $\xi \in A$ with $g'(\bar{x}, \xi) = \theta_3$ and $h(\bar{x}) + h'(\bar{x}, \xi) \geq \theta_4$ we have: There exists a function of a real variable t; V: $[0, 1] \to A$ such that

(i) $V(0) = \bar{x}$

(ii) $g(V(t)) = \theta_3$, $h(V(t)) \geq \theta_4$, $t \in [0, 1]$

(iii) $V'(t, \tau)$ exists for $t \in [0, 1]$

(iv) $V'(0, \tau) = \xi$, $\tau > 0$.

---

[44] Duboskii & Milyutin [16] utilize the fact that the set of "variations" that give the maximand value greater than the maximum could not intersect with the sets of "variations" that satisfy the constraints. By variations they mean differentials at x. Since these sets are defined by linear inequalities and equations they are convex. Using a separation theorem they derive what they call the Euler equation. Writing the Euler equation in terms of differentials of the maximand and constraints we obtain conclusion 3 of Theorem 12.

[45] For finite dimensional spaces, this is equivalent to the rank condition.

Remark 1: R.3 is a specialization of Gapushkin's regularity condition which is a uniform regularity condition[46]. R.1 is a further specialization for the case of equality constraints.

Remark 2: Recall that the inequality in constraint (2) of problem 3 is defined in terms of a closed convex cone, say, K_2. Let $K = \{\theta_3\} \otimes K_2$, where $\{\theta_3\}$ is a cone that contains only θ_3. Let $\theta = \theta_3 \otimes \theta_4$ and let $G: A \to C \otimes D$ be the "pair" valued function $< g, h >$. Then we may write constraints (1) and (2) in the form $G(x) \geq \theta$ where " $\geq$ " is in the sense of K. With that formulation, R.2 becomes a regularity condition for equality - inequality constraints.

4.1.2 Sufficient conditions for regularity.

We now present some conditions that imply regularity. We present some sufficiency lemmas for equality constraints and some sufficiency lemmas for equality-inequality constraints. These last lemmas are, of course, sufficiency lemmas, for R.1 and R.1 regularity. However, they may be strengthened by specializing the conditions when we are concerned with R.1 - regularity or R.2 - regularity. Before stating these conditions we introduce some notations.[47]

0.1) The constraint set $N = \{x \in A: g(x) = \theta_3 \text{ and } h(x) \geq \theta_4\}$.

0.2) Let $|| \cdot ||_A$ be the norm of the space A, a sphere in A with center at $\bar{x}$ and radius δ will be denoted by $\gamma(\bar{x}, \delta)$ and $\gamma(\bar{x}, \delta) = \{x: || x - \bar{x} ||_A \leq \delta\}$.

0.3) The set $D_{\bar{x}} = \{\xi \in A: g'(\bar{x}, \xi) = \theta_3\}$ is a subspace of A. Let $P_{\bar{x}}$ be the projection operator with $P_{\bar{x}}A = D_{\bar{x}}$.

0.4) Let Z be a linear space, we denote by Z^* the space of line are functionals defined on Z.

0.5) Denote by N_δ the set $N_\delta = \bigcup_{x_o \in N} \gamma(x_o, \xi)$.

We now list the conditions which we use in the statements of sufficiency theorems for regularity.

S.1 The function $h'(x, \xi)$ and $g'(x, \xi)$ are continuous and bounded on N.

---

[46] Section 2, page 592, in the sense that the condition holds for all points of the constraint set.

[47] This is Gapushkin's [21] notation.

S.2 $g'(x, \xi)$ maps A onto C. Furthermore, the space A may be written as the direct sum of $D_{\bar{x}}$ and another subspace $E_{\bar{x}}$, i.e., $A = D_{\bar{x}} \oplus E_{\bar{x}}$ where the projection operator (see 0.3) $P_{\bar{x}}$ is bounded, i.e., there exists a positive constant M such that $\| P_{\bar{x}} \|_A \leq M$.

S.2' The space A^* may be written as the direct sum of two subspaces $R^*_{\bar{x}}$ and $S^*_{\bar{x}}$ i.e., $A^* = R^*_{\bar{x}} \oplus S^*_{\bar{x}}$, where $R^*_{\bar{x}} = \{g'^*(\bar{x}, \xi^*_1) = 0, \xi^*_1 \in C^*\}$ and where g'^* is the conjugate[48] operator of g' and where $S^*_{\bar{x}}$ is a subspace of A^*. Furthermore, the projection operator $Q_{\bar{x}}$ with $Q_{\bar{x}}A^* = R^*_{\bar{x}}$ is bounded.

S.3 Given $\delta > 0$. For any $x \in \gamma(\bar{x}, \delta)\cap N$, the space A can be written as the direct sum of $A = D_x \oplus E_x$ and the projection operators P_x with $P_x A^* = R_x$ are bounded and satisfy the Lipschitz condition, i.e., $\| P_x \| \leq M$ and $\| P_{x_1} - P_{x_2} \|_A \leq M_1 \|x_1 - x_2 \|_A$ for $x, x_1, x_2 \in \gamma(\bar{x}, \delta)\cap N$, where M and M_1 are positive constants.

S.3' For any $x \in \delta(\bar{x}, \delta)\cap N$ we have $A^* = R^*_x \oplus S^*_x$ where the projection operators Q_x, with $Q_x A^* = R^*_x$, are bounded and Lipschitzian.

S.4 $\|g'^*(\bar{x}, \xi^*_1) \| \geq M \| \xi^*_1 \|$, for any $\xi^*_1 \in C^*$, where $M > 0$.

S.4' $\|g'(\bar{x}, \eta) \| \geq M_1 \| \eta \|_A$, for any $\eta \in E_{\bar{x}}$, where $M_1 > 0$.

S.4" For any $y \in C$, the equation $g'(\bar{x}, b) = y$ has a solution $b(y)$ with $\| b(y) \|_A \leq M_2 \| y \|_C$ where M_2 is a positive constant.

S.5 There exists $\bar{\xi} \in A$ with $\| \bar{\xi} \|_A \leq K$ such that: (i) $g'(x, \bar{\xi}) = \theta_3$. (ii) $[L, (h(\bar{x}) + h'(\bar{x}, \bar{\xi}))] \geqq P$, where L is a non-negative linear functional with $\| L \| = 1$ and where P and K are positive constants.

We now present sufficiency conditions for regularity. These lemmas follow from Gapushkin's theorems [21] on uniform regularity of N.

Lemma 1. <u>For equality constraints, S.3 => R.1.</u>

The lemma follows from Theorem 2 of Gapushkin [21].

Lemma 2. <u>If A is reflexive then S.3' => R.1 for equality constraints.</u>

This follows from lemma 1 (as corollary to Theorem 2 of Gapushkin [16].

Lemma 3. <u>S.1, S.2 and either S.4, S.4' or S.4" => R.1 for equality constraints.</u>

This follows from Theorem 3 of Gapushkin [21] and its corollaries.

Lemma 4. In the presence of equality and inequality constraints <u>any</u> of the following

[48] See Kantorovich [29] Chapter XII, page 476.

conditions is sufficient for R.3:

(i) The equality constraint satisfies R.1, and S.1 and S.5 are satisfied.

(ii) S.1, S.3 and S.5.

(iii) A is reflexive, S.1, S.3' and S.5.

(iv) S.1, S.2, S.4 and S.5.

(v) S.1, S.2, S.4' and S.5.

(vi) S.1, S.2, S.4" and S.5.

The lemma follows from Theorem 4 of Gapushkin [21] and from his remark at the end of section 4 of [21].

4.1.3 First order necessary conditions for the regular case.

Theorem 13. If, in addition to the assumptions of Theorem 12, $\hat{x}$ is R.3 - regular then the conclusions of Theorem 12 follow with $\lambda_o > 0$ and ℓ^1 and ℓ^2 are unique (taking $\lambda_o = 1$).

For the case of equality constraints, the theorem was proved directly[49] by Goldstine [22], utilizing S.2 without assuming that A is the direct sum of S.2.[50] The theorem was proved by Hurwicz[51] [27], and it follows from Theorem 5 of Gapushkin, who restricts A and C to be reflexive.

4.2 First order sufficient conditions.

Theorem 14. If 1) The functions f, g and h are differentiable, 2) The conclusions of Theorem 12 are satisfied, with $\lambda_o > 0$, at a point $\hat{x}$ that satisfies the constraints of problem 3, and if either: 3.a) The functional F of Theorem 12 is concave, 3.b) The equality constraint g is linear and f and h are concave. Then $\hat{x}$ is a global solution of problem 3.

For an outline of the proof of this theorem see the proof of theorem V.3.3. of Hurwicz [27] where he utilizes the fact that the difference between the values of a concave functional, say J(x), at two different points is less or equal to the differential, i.e., $J(x'') - J(x') \leq J'(x'' - x'))$. Guignard [23], using a constraint

---

[49] Without using Theorem 12, Theorem 2.1.

[50] See Liusternik and Sobolev [34], (page 204) for a proof that this part of S.2 is dispensable and for an elegant proof of Theorem 13 for equality constraints.

[51] Theorem V.3.3.2 (page 97), see remark 1 in 4.1.1 of this chapter.

qualification, proves Theorem 14 with pseudo-concavity of f and h replacing assumption 3 of the theorem (in the absence of equality constraints).

4.3 Second order necessary conditions.

Conjecture 1. If 1) the functions f, g, and h have second order differentials, 2) $\hat{x}$ is a solution to problem 3 and 3) $\hat{x}$ is regular. Then the upper bound of $F''(\hat{x}, \xi)$ is non-positive, for ξ with $|| \xi ||_A = 1$ that satisfy a) $g'(\hat{x}, \xi) = 0$, b) If h is effective i.e., if $h(\hat{x}) = 0$ then $h'(\hat{x}, \xi) = 0$, where F is as defined in Theorem 12.

For equality constraints the conjecture was proved by Goldstein[52] [22].

4.4 Second order sufficient conditions.

Conjecture 2. If 1) f, g and h have second differential, 2) The conclusions of Theorem 12 are satisfied at a point $\hat{x}$ that satisfies constraints 1) and 2) of problem 3, 3) The point $\hat{x}$ is regular and 4) The upper bound of $F''(\hat{x}, \xi)$ is negative for with $|| \xi || = 1$ satisfying a) $g'(\hat{x}, \xi) = 0$ and b) if h is effective at $\hat{x}$ then $h'(\hat{x}, \xi) = 0$, where F is as defined in Theorem 12. Then there exists a neighborhood $\overline{N}$ in A such that $f(\hat{x}) > f(x)$ for $x \in \overline{N} \cap N$.

This conjecture was proved by Goldstein[53] [22] for the case of equality constraints.

[52] Theorem 2.3 (page 147).

[53] Theorem 3.1 (page 148)

II

FINITE DIMENSIONAL PROBLEMS

CHAPTER 2

EQUALITY CONSTRAINTS

Let $x = (x_1, \ldots, x_n)$ be an element of E^n, the n dimensional Euclidian space, $n < \infty$, with the usual norm: $||x|| = (\sum_{i=1}^{n} x_i^2)^{1/2}$. Let f(x) be defined on all of E^n with values in R, denoting real numbers $R = E^1$. Let g(x) be defined on all of E^n with values in E^m, i.e. $g(x) = \begin{bmatrix} g^1(x) \\ g^2(x) \\ \vdots \\ g^m(x) \end{bmatrix}$ and let $g^\alpha(x)$ denote the components of g(x) as $\alpha = 1, \ldots, m$. Throughout this chapter, we shall assume that m is less than n.

Statement of the problem. In this chapter "problem A" means: maximize f(x) subject to g(x) = 0, i.e. subject to $g^\alpha(x) = 0$, $\alpha = 1, \ldots, m$. We now define two types of solutions to the problem: A local solution to problem A is a point $\hat{x}$ with $g(\hat{x}) = 0$ such that $f(\hat{x}) \geq f(x)$ for x in E^n satisfying g(x) = 0 provided x is in some ε-neighborhood of $\hat{x}$, $N(\hat{x}, \varepsilon)$, i.e. x is an element of the set $N(\hat{x}, \varepsilon) = \{x : ||x - \hat{x}|| \leq \varepsilon\}$, for some $\varepsilon > 0$. If $N(\hat{x}, \varepsilon) = E^n$, e.g. if $\varepsilon = \infty$, then $\hat{x}$ is called a global solution to problem A. It is clear from the definitions that a global solution is also a local solution and that the converse is not true.

In this chapter we characterize solutions to problem A by way of derivatives. We use the following notation for derivatives: Let G(x, y) be a function from $E^{n_1 + n_2}$ to E^{n_3}, i.e. x has n_1 components, y has n_2 components and G has n_3 components. We shall say that G^j, a given component function of G, is of class $C^{(r)}$ in x (in y) if it is continuous and passes continuous partial derivatives in x (in y) of order less than or equal to r. Let all components of G be of class C^1 in x and y. By G_x we denote the $n_3 \times n_1$ matrix whose components are the partial derivatives $G^j_{x_i} = \frac{\partial G^j}{\partial x_i}$, $i = 1, \ldots, n_1$, $j = 1, \ldots, n_3$, and G_y denotes the $n_3 \times n_2$ matrix whose components are the partial derivatives $G^j_{y_k} = \frac{\partial G^j}{\partial y_k}$, $k = 1, \ldots, n_2$, $j = 1, \ldots, n_3$. In case we are speaking of derivatives evaluated at a point (x_o, y_o) we denote the fact by $\overset{o}{G}_x$ and $\overset{o}{G}_y$. We also introduce the notation for second derivatives of a scalar valued function, for a particular component function

G^j of class C^2. G^j_{xx} denotes the $n_1 \times n_1$ matrix with components $\frac{\partial^2 G^j}{\partial x_i \partial x_k}$, where $i, k = 1, \ldots, n_1$, G^j_{xy} denotes the $n_1 \times n_2$ matrix with components $\frac{\partial^2 G^j}{\partial x_i \partial y_\ell}$, $i = 1, \ldots, n_1$, $\ell = 1, \ldots, n_2$ and G^j_{yy} denotes $\frac{\partial^2 G^j}{\partial y_r \partial y_s}$, $r, s = 1, \ldots, n_2$. In case the derivatives are evaluated at a point $(\overset{o}{x}, \overset{o}{y})$, we indicate this by $\overset{o}{G}{}^j_{xx}$, $\overset{o}{G}{}^j_{xy}$ and $\overset{o}{G}{}^j_{yy}$.

The central tool in the analysis of this chapter is the implicit function theorem, which we now state.

An implicit function theorem: Consider the function $G(x, y)$ defined above. If 1) $n_1 = n_3$. 2) $G(\overset{o}{v}, \overset{o}{u}) = 0$ for some point $(\overset{o}{v}, \overset{o}{u})$. 3) $G(v, u)$ is of class $C^{(r)}$ in v and of class $C^{(s)}$ and (v, u) with $r \geq 1$, $s \geq 0$. 4) The determinent $|\overset{o}{G}_v| \neq 0$. Then there exist functions $\phi(u)$ defined on some neighborhood of $\overset{o}{u}$, $N(\overset{o}{u})$, with values in E^{n_1} and a constant $\Sigma > 0$ such that:

1) $\overset{o}{v} = \phi(\overset{o}{u})$, 2) $G(\phi(u), u) = 0$ for $u \in N(\overset{o}{u})$, 3) $||\phi(u) - \overset{o}{v}|| < \Sigma$ for $u \in N(\overset{o}{u})$, 4) The functions $\phi(u)$ are of class $C^{(s)}$ on $N(\overset{o}{u})$.

For a proof see e.g., E. J. McShane and T. A. Botts [37].

1. First order necessary conditions.

In this section we show that if $\hat{x}$ is a global, hence local, solution of problem A then the partial derivatives of the "Lagrangian function", evaluated at $\hat{x}$, vanish. The Lagrangian function is defined to be a linear combination of the maximand, f, and the constraints g^α, with one as the coefficient of the maximand. This is accomplished in two steps. The first step, theorem 1, consists of proving that the derivatives of an "auxiliary Lagrangian", where the coefficient of f is a non-negative, possibly zero, constant, vanish at $\hat{x}$. The next step, theorem 2, provides a sufficient condition for the coefficient of f, in the auxiliary Lagrangian not to be zero. This sufficient condition states that the rank of the matrix $\hat{g}_x$ is maximal, i.e. equal to m. Theorem 1 is of importance by itself, for those applications where it is difficult to verify the rank condition and where it may be easy to verify directly that the coefficient of the maximand is not zero. Two examples, due to Bliss [9], that illustrate what happens when the rank condition is not satisfied are given at the end of this section.

Theorem 1. <u>If 1) f and g are of class C^1. 2) $\hat{x}$ is a global solution to problem A. Then there exists a non-zero vector $(\lambda_o, \nu) = \lambda_o, \nu^1, \nu^2, \ldots, \nu^m)$</u> such that $\hat{F}^o_x = 0$, where $F^o = (\lambda_o, \gamma)\begin{pmatrix} f(x) \\ g(x) \end{pmatrix}$.

Proof: Consider the (m + 1) x n matrix $A = \begin{bmatrix} \hat{f}_x \\ \cdots \\ \hat{g}_x \end{bmatrix}$, i.e. the matrix of first derivatives of $\begin{pmatrix} f(x) \\ g(x) \end{pmatrix}$ evaluated at $\hat{x}$. The theorem asserts that the system:

(1) $(\lambda_o, \nu)A = 0$,

of n linear homogenous equations in (m + 1) unknowns, (λ_o, ν), has a non-trivial solution. By our hypothesis, $m < n$, we have $m + 1 \leq n$. Thus a non-trivial solution of (1) exists if and only if the rank of the matrix A is less than m + 1. So the assertion of the theorem is equivalent to the assertion: The rank of A is less than m + 1 and the theorem is proved once we show that the rank of A is less than m + 1. We shall do that by contradiction; we show that if the rank of A is greater than or equal to m + 1, then $\hat{x}$ is not a local, hence is not a global, solution to our problem. Thus contradicting our assumption. Now, the contradiction hypothesis and the hypothesis that $m + 1 \leq n$ imply that the rank of A is m + 1 ($m + 1 \leq n$ implies that the rank of A is $\leq m + 1$ and the contradiction hypothesis imply that the rank of A is $\geq m + 1$), i.e. that A has m + 1 linearly independent columns. Renumber the components of x so that the first m + 1 components of A are those linearly independent columns. Let the first m + 1 components of, the rearranged, x be denoted by x_I. Let the "other" components of x be denoted by x_{II}. Let $y = (x_{II}, p)$ where p is a real variable.

Now consider the function $G(x_I, y)$ defined on $E^{m+1, n-(m+1)+1}$ with values in E^{m+1}, where the component functions of G are:

$$G^o(x_I, y) = f(x_I, x_{II}) - f(\hat{x}_I, \hat{x}_{II}) - p$$

$$(2) \quad G^1(x_I, y) = g^i(x_I, x_{II})$$

$$\vdots$$

$$G^m(x_I, y) = g^2(x_I, x_{II})$$

We shall apply the implicit function theorem to $G(x_I, y)$ with "initial" point

$(\hat{x}_I, \hat{y}) = (\hat{x}_I, \hat{x}_{II}, 0)$. If we could do that then we would obtain a solution $x_I = \zeta_I(y)$ so that $G(\zeta_I(y), y) = 0$ in some neighborhood of $(x_I, \hat{x}_{II}, p) = (\hat{x}, 0)$. Thus, in that neighborhood, the last m component functions of (2) equal zero, i.e. the constraints are satisfied. And the first component function of (2) also equals zero, i.e. $f(x) - f(\hat{x}) =$ for p in some neighborhood of $p = 0$. Picking $p > 0$, we would find a point $\overline{x}$ that satisfies the constraints and such that $f(\overline{x}) > f(\hat{x})$. This, then, would complete our proof, so all that remains now is to show that the implicit function theorem applies.

Refering to the implicit function theorem stated at the beginning of this chapter, we verify the conditions of that theorem. Take $v = x_I$, $u = y$, $n_1 = n_3 = m + 1$, $n_2 = n - m$ and $(\overset{o}{u}, \overset{o}{v}) = (\hat{x}_I, \hat{x}_{II}, 0)$. Conditions 1) and 2) of the theorem are satisfied by our choice of variables and by direct substitution in 2) since $\hat{x}$ satisfies the constraints. Condition 3) is satisfied with $r = s = 1$ by assumption 1) of the present theorem and by linearily in p. Condition 4) of the theorem is satisfied since $|\overset{o}{G}_v| = |\hat{G}_{x_I}| \neq 0$ by the contradiction hypothesis and by the choice of the components of x_I. Q.E.D.

Theorem 2. <u>If in addition to the assumptions of theorem 1--the rank of the matrix $[\hat{g}_x]$ is equal to m, then the conclusion of theorem 1 holds with $\lambda_o > 0$, i.e. there exists a unique vector $\lambda = (\lambda^1, \ldots, \lambda^m)$ such that $\hat{F}_x = 0$, where $F = f(x) + \lambda g(x)$.</u>

Proof: Suppose not, i.e. suppose $\lambda_o = 0$. Then, by theorem 1, there exists a non-zero vector ν such that $\hat{F}^o_x = \nu\hat{g}_x = 0$. But this means that the system, of n^1 linear homogeneous equations in m unknowns, $\nu\hat{g}_x = 0$ has a non-trivial solution ν. This, in turn, implies that the rank of $\hat{g}_x$ is less than m, since $m < n$. Thus, by contradiction, $\lambda_o > 0$. Now define $\lambda^\alpha = \frac{\nu^\alpha}{\lambda_o}$. By theorem 1, dividing both sides of $\hat{F}^o_x = 0$ by λ_o, we have: there exists a vector λ such that $\hat{F}_x = 0$. It remains to show that λ is unique. Suppose $\lambda' \neq \lambda$ satisfies $\hat{F}'_x = 0$, where $\hat{F}' = f(x) + \lambda' g(x)$. Then $\hat{F}_x - \hat{F}'_x = (\lambda - \lambda')\hat{g}_x = 0$, i.e. there exists a non-zero vector $(\lambda - \lambda')$ satisfying the system $(\lambda - \lambda')\hat{g}_x = 0$. This again contradicts the assumption that $\hat{g}_x$ has, maximal, rank m. Q.E.D.

Examples:

1) Bliss [9], Suppose $n = 2$, $m = 1$ and suppose $f(x_1, x_2)$ has non-zero derivatives at $x = (0, 0)$. Let $g(x) = x_1^2 + x_2^2$. The problem, then, is to maximize $f(x)$ subject to $x_1^2 + x_2^2 = 0$. The only point in the constraint set is $(0, 0)$ and the maximum must occur there. By theorem 1 we have $\lambda_o \hat{f}_{x_1} + 2\lambda\hat{x}_1 = 0$ and $\lambda_o \hat{f}_{x_2} + 2\lambda\hat{x}_2 = 0$, $(\lambda_o, \lambda) \neq 0$. Thus $\lambda_o f_{x_1}(0, 0) = 0$ and $\lambda_o f_{x_2}(0, 0) = 0$ with λ arbitrary e.g. $\lambda = 1$. Since $\hat{f}_{x_1} \neq 0$ and $\hat{f}_{x_2} \neq 0$ we must have $\lambda_o = 0$. The matrix $\hat{g}_x = (2\hat{x}_1 \quad 2\hat{x}_2) = (0, 0)$ and has rank zero and theorem 2 does not apply.

2) (Bliss [9]) $m = 2$, $n = 1$, $f = -x_2^2 - 2\,x_1^2$, $g = x_1^2x_2 - x_2^3$. The point that maximizes f subject to $x_1^2x_2 - x_2^3 = 0$ is again $(0, 0)$. For the constraint set is given by $\{(x_1, x_2) \mid x_2 \neq 0 \text{ and } x_1 = x_2\} \cup \{(x_1, x_2) \mid x_2 = 0\}$. On the first set $f = -x_2$ and on the second set $f = -2x_1$. Again, theorem 1 applies, and $-4\lambda_o x_1 + 2\lambda\hat{x}_1\hat{x}_2 = 0$, $2\lambda_o\hat{x}_2 + \lambda(\hat{x}_1^2 - 3\hat{x}_2^2) = 0$, $(\lambda_o, \lambda) \neq 0$. The conclusion of theorem 1 could be verified, at $(0, 0)$, by an arbitraty choice of (λ_o, λ). The rank of $\hat{g}_x$ = rank of $(2\hat{x}_1\hat{x}_2 - 3\hat{x}_2^2)$ = rank of $(0, 0) = 0$, and we can not apply theorem 2. But the conclusion of theorem 2 holds.

2. Second order necessary conditions.

In the section we show that a necessary condition for $\hat{x}$ to solve problem A is that the second differential of the Lagrangian function, $\eta\hat{F}_{xx}\eta^*$, be non-positive for η satisfying $\hat{g}_x\eta^* = 0$. The method of proof is to parametrize the constraint set, i.e. express the point x that satisfies $g(x) = 0$ as functions of a real variable, and then apply the second order necessary condition for a nonconstrained maximum of a function of one real variable. We first state theorem 3, then a parametrization lemma which we use in the proof of theorem 3. Then we prove the lemma.

Theorem 3. If 1) f and g are of class C^2. 2) $\hat{x}$ is a global solution to problem A. 3) the rank of $[\hat{g}_x]$ is m. Then $\eta\hat{F}_{xx}\eta^* \leq 0$ for all η with $\hat{g}_x\eta^* = 0$, where "$*$" denotes the transpose.

Lemma: (parametrizability). If conditions 1) and 3) of theorem 3 hold and if $\eta = x - \hat{x}$ satisfies $\hat{g}_x\eta^* = 0$, where $g(\hat{x}) = 0$, then there exists an n-vector valued function $\omega(b)$ of a real variable b such that:

(1) $\omega(b)$ is of class C^2 in a neighborhood S of $b = 0$.

(2) $g[\omega(b)] = 0$ for $b \varepsilon S$, i.e. if we substitute $x = \omega(b)$ in $g(x)$ then the constraints are satisfied for all $b \varepsilon S$.

(3.a) $\omega(0) = \hat{x}$.

(3.b) $\omega'(0) = \eta$, where "'" denotes the derivative.

Proof of theorem 3: Let η satisfy $\hat{g}_x\eta^* = 0$. We wish to show that $\eta\hat{F}_{xx}\eta^* \leq 0$. By the lemma, we can write $x = \omega(b)$ for $b \varepsilon S$ where $\omega(0) = \hat{x}$ and $g[\omega(b)] = 0$. Thus, $b = 0$, by hypothesis of the theorem, provides a local unconstrained maximum for $\phi(b) = f[\omega(b)]$ in S. Hence:

(i) $\phi'(0) = 0$.

(ii) $\phi''(0) \leq 0$.

But, using the chain rule,

(iii) $\phi'(b) = \sum_{i=1}^{n} f_i[\omega(b)]\omega_i'(b)$.

Now, consider Lagrangian function F whose derivatives with respect to x are the components of the vector:

(iv-1) $F_x = f_x + \lambda g_x$

Post-multiplying both sides of (iv-1) by $(\omega'(b))^*$, i.e. by the column vector whose components are $\omega_i'(b)$, and substituting $x = \omega(b)$ we have:

(iv-2) $F_x(\omega')^* = f_x(\omega')^* + \lambda g_x(\omega')^*$.

By lemma 1, $g[\omega(b)] = 0$ for $b \varepsilon S$. Thus $\frac{d}{db} g[\omega(b)] = g_x(\omega'(b))^* = 0$ for $b \varepsilon S$. Premultiplying by λ we have:

(v) $\lambda g_x(\omega'(b))^* = 0$, $b \varepsilon S$.

By (iv-2) and (v) we have: $F_x(\omega') = f_x(\omega')^*$, thus, by (iii), $\phi'(b) = F_x(\omega'(b))^*$. Differentiating the last expression with respect to b we get: $\phi''(b) = \omega'(b)F_{xx}(\omega'(b))^* + F_x(\omega''(b))^*$, where $(\omega''(b))^*$ is the n-column vector whose components are the second derivatives $\omega''(b)$.

Evaluating at $b = 0$, we get:

(vi) $\phi''(0) = \omega'(0)F_{xx}[\omega(0), \lambda](\omega'(0))^* + F_x[\omega(0), \lambda](\omega''(0))^*$.

But $F_x[\omega(0), \lambda] = \hat{F}_x = 0$, by first order necessary condition, thus the second term of the right hand side of (vi) is zero. Also, by lemma, $\omega'(0) = \eta$ and $(0) = x$. Hence we may write (vi) as:

(vii) $\phi''(0) = \eta \hat{F}_{xx} \eta^*$.

By (ii) and (vii) the proof of the theorem is complete, since η is an arbitraty vector with $\hat{g}_x \eta^* = 0$.

Proof of the lemma: Let $\eta = x - \hat{x}$ satisfy $\hat{g}_x \eta^* = 0$. By hypothesis of the lemma, the matrix g_x has rank m. Without loss of generality we may assume that the columns $\hat{g}_{x_s}$ are linearly independent, $s = 1, \ldots, m$. Let the matrix $\hat{g}_{x^I}$ denote the matrix formed by these columns. Denote by $\hat{g}_{x_\gamma}$, the remaining columns of $\hat{g}_x$ and by $\hat{g}_{x^{II}}$ the matrix formed by them. The vector x is now partioned into (x^I, x^{II}), similarly we partion η into $\eta = (\eta^I, \eta^{II})$. The assertion of the lemma is that we can express x as a function of b. Consider the n-vector function $G(x, b) = (G^I(x, b), G^{II}(x, b))$, with G^I having m components and G^{II} having n - m components, where $G^I(x, b) = g(x)$ and $G^{II}(x, b) = x^{II} - (b\eta^{II} + \eta^{II})$. We shall apply the implicit function theorem to G(x, b) with "initial point" $(x, b) = (\hat{x}, 0)$. Note that $G(\hat{x}, 0) = (g(\hat{x}), \hat{x} - \hat{x}) = 0$. Note also that the Jacobian $\hat{G}_x$ is:

$$\begin{bmatrix} \hat{g}_{x^I} & \hat{g}_{x^{II}} \\ 0 & I \end{bmatrix}$$

where 0 is an n - m x m zero matrix and I is n - m x n - m identity matrix. Clearly $\hat{G}_x$ has rank n. The functions G(x, b) are of class C^2 by hypothesis and by construction. Thus the implicit function theorem applies and we can obtain $x = \omega(b)$ for b in some neighoorhood of $b = 0$. By conclusions 1), 2) and 4) of that theorem, at the beginning of this chapter, we get conclusions (1), (2), and (3.a) of our lemma. It remains to prove (3.b). In $G^I(x, b)$ set $x = \omega(b)$. Thus $G^I(\omega(b), b) = g[\omega(b)] = 0$, $b \in S(0)$. Differentiating with respect to b we have:

(i) $\frac{d}{db} g[\omega(b)] = g_{x^I}(\omega'^I)^* + g_{x^{II}}(\omega'^{II})^* = 0.$

Now $\omega^{II}(b) = \hat{x}^{II} + b\eta^{II}$, by solving for x^{II} from $G^{II})x, b) = 0$. Thus:

(ii) $\omega'^{II}(b) = \eta^{II}$.

By (ii) we have: $\omega'^{II}(0) = \eta^{II}$, so it remains to show that $\omega'^{I}(0) = \eta^{I}$. Substitute from (ii) into (i) getting: $\frac{d}{db} g\,[\omega(b)] = g_{x^I}(\omega'^{I})^* + g_{x^{II}}(\eta^{II})^* = 0.$

Evaluating at b = 0, we get

(iii) $\hat{g}_{x^I}(\omega'^{I}(0))^* + \hat{g}_{x^{II}}(\eta^{II})^* = 0.$

Since η satisfies $\hat{g}_x\eta^* = 0$, we have:

(iv) $\hat{g}_x\eta = \hat{g}_{x^I}(\eta^I)^* + \hat{g}_{x^{II}}(\eta^{II})^* = 0.$

Subtracting (iv) from (iii) we have:

(v) $\hat{g}_{x^I}(\omega'^{I}(0) - \eta^I)^* = 0.$

But (v) is a system of linear homogeneous equations with a square nonsingular matrix of coefficients. Hence (v) can only have the trivial solution $\omega'^{I}(0) - \eta^{I} = 0$, i.e. $\omega'^{I}(0) = \eta^{I}$. Thus $\omega'(0) = \eta$ as we were to prove.

3. Second order sufficient conditions.

In this section we prove that if, in addition to the vanishing of its first derivatives, the second differential of the auxiliary Lagrangian F° is positive definite for $\eta \neq 0$ with $\hat{g}_x\eta^* = 0$ then the point $\hat{x}$ is a local solution of problem A.

Theorem 4. If 1) f and g are of class C^2. 2) $\hat{F}^{\circ}_x = 0$ at some point $\hat{x}$ with $g(\hat{x}) = 0$. 3) $\eta\hat{F}^{\circ}_{xx}\eta^* < 0$ for η satisfying $\hat{g}_x\eta^* = 0$. Then $\hat{x}$ is a local solution of problem A.

Proof: The theorem is proved by contradiction. Suppose $\hat{x}$ is not a local solution to problem A. Then there exists a sequence of distinct points $\{x_r\}$ with $x_r \neq \hat{x}$ converging to $\hat{x}$ such that $g(x_r) = 0$ and $f(x_r) \geq f(\hat{x})$. Define $k_r = ||x_r - \hat{x}||$. Then $k_r \to 0$, since $x_r \to \hat{x}$. Define $h_r^i = \frac{x_r^i - \hat{x}^i}{k_r}$, i = 1, ..., n, and let $h_r = (h_r^1, h_r^2, \ldots, h_r^n)$. Note that $||h_r|| = \frac{1}{k_r}\,||x_r|| = 1$. The sequences $\{h_r^i\}$ are bounded, hence each one of them has a convergent subsequence. We shall abuse the notation and refer to these convergent subsequences as $\{h_r^i\}$. Denote the limits of $\{h_r^i\}$ by h_0^i. Our contradiction consists of showing that $h_0^i = 0$, i.e. we have a sequence h_r with $||h_r|| = 1$ that converges to the zero vector.

Now $g(x_r) = 0$, thus $F^{\circ}(x_r) = \lambda_o f(x_r)$. By the contradiction assumption

$f(x_r) - f(\hat{x}) \geq 0$. Since $\lambda_o \geq 0$, we have $\lambda_o f(x_r) - \lambda_o f(\hat{x}) \geq 0$. But $F^o(\hat{x}) = \lambda_o f(\hat{x})$. Thus:

(1) $F^o(x_r) - F^o(\hat{x}) \geq 0$.

The elements of $\{x_r\}$ are distinct from $\hat{x}$, hence $k_r > 0$. Thus:

(2) $\frac{1}{k_r^2} (F^o(x_r) - F^o(\hat{x})) \geq 0$.

By Taylor's theorem:

(3) $F^o(x_r) - F^o(\hat{x}) = \hat{F}^o_x \eta_r^* + \frac{1}{2}\eta_r \hat{F}^o_{xx} \eta_r^* + R$,

where R is of smaller order of magnitude than $||x_r - \hat{x}||^2$ as $||x_r - \hat{x}|| \to 0$. By hypothesis $\hat{F}^o_x = 0$, hence:

(4) $F^o(x_r) - F^o(\hat{x}) = \frac{1}{2} \eta_r \hat{F}^o_{xx} \eta_r^* + R$.

Dividing both sides of (4) by k_r^2 and noting that $h_r = \frac{\eta_r}{k_r}$ we have:

(5) $\frac{1}{k_r^2} (F^o(x_r) - F^o(\hat{x})) = \frac{1}{2} h_r \hat{F}^o_{xx} h_r^* + \frac{R}{k_r^2}$

By (2) and (5) we have:

(6) $h_r \hat{F}^o_{xx} h_r^2 + 2 \frac{R}{k_r^2} \geq 0$.

Passing to the limit through the sequences $\{h_r\}$ we have:

(7) $h_0 \hat{F}^o_{xx} h_0^* \geq 0$.

We now show that $h_0 = 0$. This will be done by showing that h_0 satisfies $\hat{g}_x h_0^* = 0$, for in that case either $h_0 = 0$ or $h_0 \hat{F}_{xx} h_0^* < 0$ with the later possibility contradicting (7). Since $g(x_r) = g(\hat{x}) = 0$, we have $\frac{1}{k_r} [g(x_r) - g(\hat{x})] = 0$. By the mean value theorem we have:

(8) $0 = \frac{1}{k_r} [g^\alpha(x_r) - g^\alpha(\hat{x})] = \frac{1}{k_r} g^\alpha_x(\hat{x} + \theta\eta_r)\eta_r^* = g^\alpha_x(\hat{x} + \theta\eta_r)h_r^*$.

By continuity of g^α_x, taking the limit as the subsequences $\{h_r\} \to \{h_0\}$ we have, since $\hat{x} + \theta\eta_r \to \hat{x}$,

(9) $\hat{g}_x h_0^* = 0$.

By (9) the proof is complete.

CHAPTER 3

INEQUALITIES AS ADDED CONSTRAINTS

Let h(x) be defined on E^n with values in E^ℓ, i.e., h(x) is a column vector with ℓ components. Denote the components of h by h^β, $\beta = 1, \ldots, \ell$.

Statement of the Problem. In this chapter, problem B means: maximize f(x) subject to g(x) = 0 and to $h(x) \geq 0$, where f and g are as in Chapter 2.

We might note that problem A is a special case of problem B, by setting $h'(x) = x - x$ in problem B. Another special case is where we have only inequality constraints, i.e. where g(x) = x - x, we shall refer to this as problem C. We prove first order conditions for problem B (theorem 1), first order sufficient conditions for problem C(theorem 2), second order necessary conditions for problem B (theorem 3) and second order sufficient conditions for problem B (theorem 4).

1. First order necessary conditions.

We start by introducing some notation and definitions. Let $\hat{x}$ be a point in E^n. We say that h^β is effective at $\hat{x}$ if $h^\beta(\hat{x}) = 0$. For a given $\hat{x}$, let the number of effective constraints be a. We shall remember the constraints so that the first a components of h are effective. Let h^I be the vector of effective constraints, at $\hat{x}$, and let h^{II} denote the rest of the constraints. Thus, $h(\hat{x}) = (h^I(\hat{x}), h^{II}(\hat{x}))$. Finally, let I_z denote a diagonal matrix of order ℓ with z^β's on the diagonal. Similarly define I_{z^I} and $I_{z^{II}}$ to be such matrices of order a and ℓ - a respectively.

Theorem 1. If 1) f, g and h are of class C^2, 2) $\hat{x}$ is a local solution to problem B, 3) the matrix $\begin{pmatrix} \hat{g}_x \\ \hat{h}^I_x \end{pmatrix}$ has rank m + a n. Then there exists a unique vector (λ, μ) such that:

(a) $\hat{L}_x = 0$, where $L = f(x) + \lambda g + \mu h$.

(b) $\mu^\beta h^\beta(\hat{x}) = 0, \quad \beta = 1, \ldots, \ell$.

(c) $\mu^\beta \geq 0, \quad \beta = 1, \ldots, \ell$.

Proof: We may rewrite the inequalities $h^\beta \geq 0$ as:

1) $h^\beta(x, z) = h^\beta(x) - (z^\beta)^2 = 0$.

Thus, by hypothesis, $(\hat{x}, \hat{z})$ is a local solution of the following problem of type A:

Maximize $f(x)$ subject to $g(x) = 0$, $H(x, z) = 0$. We shall apply theorem 2.2[†] and we presently show that we may. First we note that the differentiability conditions of theorem 2.2 are satisfied in view of assumption 1 of the present theorem. We now check the rank condition. We must show that the rank $J = \begin{pmatrix} \hat{g}_x & \hat{g}_z \\ \hat{H}_x & \hat{H}_z \end{pmatrix}$ is $m + \ell$. By hypothesis we have $J^1 = \begin{pmatrix} \hat{g}_x \\ \hat{h}^I_x \end{pmatrix}$ has rank $m + a$. Renumbering the variables so that the first $m + a$ components of x correspond to linearly independent columns of J^1, and partioning x accordingly in x^I and x^{II} we may write J^1 as $\begin{pmatrix} \hat{g}_{x^I} & \hat{g}_{x^{II}} \\ \hat{h}^I_{x^I} & \hat{h}^I_{x^{II}} \end{pmatrix}$. Let $\hat{z}^I$ be the vector of components of $\hat{z}$ corresponding to h^I and let $\hat{z}^{II}$ denote the remaining components of $\hat{z}$. We note that $\hat{z}^I = 0$, for $H^I = h^I$ at $\hat{x}$, and that all of the components of $\hat{z}^{II}$ are nonzero. Now $\hat{g}_{z^I} = 0_1$, where 0_1 is an $m \times a$ zero matrix and $\hat{g}_{z^{II}} = 0_2$, where 0_2 is an $m \times \ell - a$ zero matrix. $\hat{H}^I_{z^{II}} = -2I_{\hat{z}^I}$, where $I_{\hat{z}^I} = I_{z^I} \big|_{z^I = \hat{z}^I}$. But $z^I = 0$, so that $\hat{H}^I_{z^I} = 0_3$, where 0_3 is an $a \times a$ zero matrix. $\hat{H}^I_{z^{II}} = 0_4$, where 0_4 is an $a \times \ell - a$ zero matrix. $H^{II}_{z^I} = 0_5$, where 0_5 is an $\ell - a \times a$ zero matrix and $H^{II}_{z^{II}} = -2I_{\hat{z}^{II}}$, where $I_{z^{II}} = I_{z^{II}} \big|_{z^{II} = \hat{z}^{II}}$. We now write J as:

$$\begin{bmatrix} \hat{g}_{x^I} & \hat{g}_{x^{II}} & \hat{g}_{z^I} & \hat{g}_{z^{II}} \\ \hat{H}^I_{x^I} & \hat{H}^I_{x^{II}} & \hat{H}^I_{z^I} & \hat{H}^I_{z^{II}} \\ \hat{H}^{II}_{x^I} & \hat{H}^{II}_{x^{II}} & \hat{H}^{II}_{z^I} & \hat{H}^{II}_{z^{II}} \end{bmatrix} = \begin{bmatrix} \hat{g}_{x^I} & \hat{g}_{x^{II}} & 0_1 & 0_2 \\ \hat{h}^I_{x^I} & \hat{h}^I_{x^{II}} & 0_3 & 0_4 \\ \hat{h}^{II}_{x^I} & \hat{h}^{II}_{x^{II}} & 0_5 & -2I_{\hat{z}^{II}} \end{bmatrix}$$

It is clear that if [*] $m + \ell < n$, then J has rank $m + \ell$ since the square submatrix

$$\begin{bmatrix} \hat{g}^I_{x^I} & 0_2 \\ \hat{h}^I_{x^I} & 0_4 \\ \hat{h}^{II}_{x^I} & -2I_{\hat{z}^{II}} \end{bmatrix}$$

of order $m + \ell$ is non-singular. Now, theorem 2.2 applies. We get: There exists a unique vector (λ, μ) such that $\hat{F}_x = 0$ and $\hat{F}_z = 0$, where

(*) If $m + \ell \geq n$, but $m + a < n$, the theorem still holds since ineffective constraints may be, locally, ignored.

(†) Theorem 2 of chapter 2.

$F = f(x) + \lambda g + \mu H$. Writing out these conditions we have:

(1) $\hat{F}_x = \hat{f}_x + \lambda \hat{g}_x + \mu \hat{h}_x = \hat{L}_x = 0.$

(2) $\hat{F}_{z^\beta} = -2\mu^\beta \hat{z}^\beta = 0$, i.e. $\mu^\beta \hat{z}^\beta = 0$.

By (1) we have conclusion (a) of the present theorem. (b) follows from (2) since $\hat{z}^\beta \neq 0$ if and only if $h^\beta(\hat{x}) > 0$. We now show that $\mu^I \geq 0$. For the ineffective constraints, h^{II}, $\mu^{II} = 0$. It then remains to show that $\mu^I \geq 0$. In view of the continuity of h^{II}, the components of h^{II} will remain positive in some neighborhood of $\hat{x}$. In that neighborhood $\hat{x}$ is a local solution of the problem: max $f(x, z) = f(x)$ subject to $g(x) = 0$ and $H^I(x) = 0$, where $H^I(\hat{x}) = h^I(\hat{x}) = 0$. The conditions for theorem 2.3 (second order necessary conditions) are satisfied. Thus, we have:

(3) $\eta \hat{F}'_{xx} \eta^* + 2\eta \hat{F}'_{xz} \zeta^* + \zeta F'_{zz} \zeta^* \leq 0$, where $F' = f + \lambda g + \mu^I H^I$, provided:

(4.1) $\hat{g}_x \eta^* + \hat{g}_z \zeta^* = g_x \eta^* = 0$, and

(4.2) $\hat{H}^I_x \eta^* + \hat{H}^I_z \zeta^* = \hat{h}^I_x \eta^* - 2I_{z^I} \zeta^* = \hat{h}^I_x \eta^* = 0$, since $\hat{z}^I = 0$.

(3) may be written as:

(5) $\eta \hat{F}'_{xx} \eta^* - 2\mu^I I_\zeta \zeta^* \leq 0$, with η, ζ satisfying (4), since $\hat{F}'_{xz} = 0$, where I_ζ is a diagonal matrix of order a with ζ's on the diagonal. But (4) does not restrict ζ. Take $(\bar{\eta}; \bar{\zeta})$ to be a vector all of whose components are zero's except for one component of ζ, say $\zeta^{\bar{\beta}}$. $(\bar{\eta}; \bar{\zeta})$ satisfies (4). Thus, (5) holds. Hence $-2\mu^{\bar{\beta}} \bar{\zeta}^{\bar{\beta}^2} \leq 0$. Thus, $\mu^{\bar{\beta}} \geq 0$. Since $\bar{\beta}$ is arbitrary, conclusion C of the theorem is proved. This completes the proof of the theorem.

2. <u>First order sufficient conditions.</u>

In this section we show that, in the case where the equality constraints are linear, if the maximand and inequality constraints are concave, then the vanishing of the first derivatives of the Lagrangian at x is sufficient for x to be a solution to problem B. For the sake of completeness we provide differential characterizations of concavity.

Lemma: <u>If the real valued $\phi(x)$ is of class C^2 then the following statements are equivalent for $x \in E^n$.</u>

<u>(i) $\phi(x)$ is concave.</u>

<u>(ii) $\phi(x) - \phi(x^o) \leq \phi^o_x \xi^*$, $\xi = x - x^o$.</u>

(iii) $\xi\phi^o_{xx}\xi^* \leq 0$ for all ξ in E^n.

Remark: In the statement of the lemma, E^n may be replaced by a convex subset of E^n.

The proof of the lemma may be found e.g. in Fleming [20]*.

Theorem 2. If $g(x) = Ax^* + b$, where A is an m x n matrix, f(x) and h(x) are concave and of class C^2 and if: a) $g(x) = 0$, $h(x) \geq 0$, b) There exists a vector $(\lambda_o, \lambda, \mu)$ such that:

(b.i.) $\mu \geq 0$, $\mu g(\hat{x}) = 0$, $\lambda_o > 0$.

(b. ii.) $\hat{L}^o_x = 0$, where $L^o = \lambda_o f + \lambda g + \mu h$.

Then $\hat{x}$ is a global solution of problem B.

Proof: L^o is concave, since λg is linear in x, μh is a non-negative linear combination of concave functions and since $\lambda_o f$ is concave. Thus, by lemma:

$L^o(x) - L^o(\hat{x}) \leq \hat{L}^o_x\xi = 0$, for all x. But, $L^o(\hat{x}) = \lambda_o f(\hat{x})$ and for x satisfying the constraints $L^o(x) = \lambda_o f(x) + \mu h(x) \geq \lambda_o f(x)$. Thus, $\lambda_o(f(x) - f(\hat{x})) \leq L^o(x) - L^o(\hat{x}) \leq 0$. Since $\lambda_o > 0$, $f(x) \leq f(\hat{x})$ for all x satisfying $g(x) = 0$ and $h(x) \geq 0$, and the proof is complete.

3) Second order necessary conditions.

In this section we utilize theorem 2.3 to derive a second order necessary condition for problem B.

Theorem 3. If the assumptions of theorem 1 are satisfied, then $\eta\hat{L}_{xx}\eta^* \leq 0$, for η satisfying: a) $g_x\eta^* = 0$, b) $\hat{h}^I_x\eta^* = 0$, where h^I is the vector of effective constraints at $\hat{x}$.

Proof: As we did towards the end of the proof of theorem 1 of this chapter, we may express the assumption that $\hat{x}$ is a local solution of problem B as follows: $f(\hat{x}, \hat{z}^I) = f(\hat{x}, 0) = f(\hat{x})$ is a local maximum of $f(x, z) = f(x)$ in some neighborhood of $(\hat{x}, 0)$ satisfying: $g(x, z) = g(x) = 0$, $H^I(x, z) = h^I(x) - I_{z^I}z^{I*} = 0$. As we noted in the proof of theorem 1, theorem 2.3 applies and we have:

$\eta\hat{F}'_{xx} - 2\mu^I I_\zeta\zeta^* \leq 0$ provided:

1) $\hat{g}_x\eta^* = 0$.

* See also lemmas 3 and 4 of chapter 4 of these notes.

2) $\hat{h}_x^I \eta^* = 0$.

Since $\hat{F}'_{xx} = \hat{L}_{xx}$ and since 1) and 2) impose no restrictions on ζ, we get:

3) $\eta \hat{L}_{xx} \eta^* - 2\mu^I \zeta I_1 \zeta^* \leq 0$ for all ζ and for η satisfying 1) and 2). Hence, in particular for $\zeta = 0$ we have: $\eta \hat{L}_{xx} \eta^* \leq 0$ for η satisfying 1) and 2). This completes our proof.

4) Second order sufficient conditions.

In this section we use theorem 2.4 to obtain a sufficient condition for a point to be a solution of problem B.

Theorem 4. If 1) f, g and h are of class C^2, 2) the point $\hat{x}$ is such that $g(\hat{x}) = 0$ and $h(\hat{x}) \geq 0$, 3) there exists a vector $(\lambda_o; \lambda; \mu)$ such that: 3.a) $\hat{L}^o_x = 0$, where $L^o = \lambda_o f + \lambda g + \mu h$, and 3.b) $\mu \geq 0$, $\mu^\beta g^\beta(\hat{x}) = 0$. 4) $\eta \hat{L}_{xx} \eta^* < 0$ for η satisfying $\hat{g}_x \eta^* = 0$, $\hat{h}^I_x \eta^* = 0$. Then $\hat{x}$ is a local solution of problem B.

Proof: We wish to show that there exists a neighborhood of $\hat{x}$, $s(\hat{x})$, such that $f(\hat{x}) - f(x) > 0$ for all x with $g(x) = 0$ and $h(x) \geq 0$. Since $h^{II}(x)$ are continuous and $h^{II}(\hat{x}) > 0$, there exists a neighborhood $T(\hat{x})$ such that $h^{II}(x) > 0$ for $x \in T(\hat{x})$. It suffices to show that there exists a neighborhood $S^1(\hat{x})$ such that $f(\hat{x}) \geq f(x)$ for $x \in S^1(\hat{x})$ satisfying $g(x) = 0$ and $h^I(x) \geq 0$, for we may take $S(\hat{x})$ to be $S(\hat{x}) \cap T(\hat{x})$. As we did in the proofs of theorems 1 and 2, we may write $h^I(x) \geq 0$ as $H^I(x, z) = h^I(x) - z^I I_1 z^{I*} = 0$, where $\hat{z}^I = 0$, and $g(x) = 0$ as $g(x, z) = g(x) = 0$. The proof is complete if we show that there is a neighborhood $S^2(\hat{x}, 0)$ such that $(\hat{x}, 0)$ maximizes $f(s, z) = f(x)$ among $(x, z) \in S^2$ satisfying the above equality constraints. To do that we show that theorem 2.4 applies. By assumption 1) of this theorem and by definition of H, assumption 1) of theorem 2.4 is satisfied. Note that $F^o = \lambda_o f + \lambda g + \mu^I H^I = L^o - \mu^I I_{z^I} z^{I*}$. Thus, $\hat{L}^o_x = 0$ implies $\hat{F}^o_x = 0$, since $\hat{z}^I = 0$. Let $Q_o = (\eta, \zeta) \hat{F}^o_{xx} (\eta, \zeta)^*$ and let $Q_1 = \eta \hat{L}^o_{xx} \eta^*$. We wish to show that $Q_o < 0$ for (η, ζ) with:

(i.a.) $\hat{g}^I_x \eta^* + \hat{g}_{z^I} \zeta^* = \hat{g}_x \eta^* = 0$.

(i.b.) $\hat{H}^I_x \eta^* + \hat{H}^I_z \zeta^* = \hat{h}^I_x \eta^* - 2I_{z^I} \zeta^* = \hat{h}^I_z \eta^* = 0$.

By hypothesis, $Q_1 < 0$ subject to (i). But $Q_o = Q_1 - 2\mu^I I_\zeta \zeta^* \leq Q_1$ for all ζ since $\mu^I \geq 0$ and $\mu I_\zeta \zeta^* = \Sigma \mu^\beta (\zeta^\beta)^2$. Thus, $Q_o < 0$ subject to (i) and theorem 3.4 applies, as we were to show.

CHAPTER 4

EXTENSIONS AND APPLICATIONS

In this chapter we extend the results of Chapter 3 to deal with the case of vector maxima. We then relate the results of Chapter 3 to saddle value problems. As we have pointed out, no new results are presented. But we have a unified and, hopefully, more direct treatment of the problems.

In Section 1 we take a special case of the problem of Chapter 3, the case where the constraints are only of the inequality type, and relate the results obtained there to saddle value problems. This results in an alternative Kuhn-Tucker [32] equivalence theorem.

In Section 2 we extend some results of Chapter 3 and Section 1 to vector maxima. In section 3 we provide an introduction to discrete time control problems.

The first part of Section 4 is devoted to applying theorems 1 and 3 of Chapter 3 to the theory of consumer's optimum. In the second part of Section 4 we study the relation between Pureto optimality and competitive equilibrium. The proof is a straightforward application of the results of Section 2.

1. Relation to Saddle Value Problems.

Here the constraints are of the inequality type and they involve non-negativity constraints of the variables. We show, theorem 1, that a local constrained maximum implies the existence of a non-negative saddle point under certain conditions. And conversely, theorem 2, that a non-negative saddle point implies a local constrained maximum. Corollary 1 of theorem 1 and theorem 2 is a restatement of theorem 3 of Kuhn and Tucker [32]. We then relate theorem 2, which is a sufficiency theorem, to the sufficiency of Chapter 3.

Theorem 1 differs from the necessity part of Kuhn and Tucker theorem 3 [22] in the proof of the quasi-saddle point conditions and in the form of the constraint qualification. We use a form of the constraint qualification presented in the Arrow-Hurwicz-Uzawa paper [3] on the subject.

For completeness we restate lemmas 1 and 2 of Kuhn and Tucker together with their sufficient conditions for the hypothesis of lemma 2 of theirs.

We now define a non-negative saddle value of a function. Consider the function $G(x, y)$, where x is a vector with n components and where y is a vector with m

Definition: A point $(\hat{x}, \hat{y}) \geq 0$ is said to be a saddle value to $G(x, y)$ if:

1.a) $G(x, \hat{y}) \leq G(\hat{x}, \hat{y})$ for all $x \geq 0$.

1.b) $G(\hat{x}, \hat{y}) \leq G(\hat{x}, y)$ for all $y \geq 0$.

We call $(\hat{x}, \hat{y})$ a non-negative saddle point.

Let us now present two preliminary lemmas relating saddle values to the derivatives of $G(x, y)$.

Lemma 1). If

1) $G(x, y)$ is differentiable,

2) $(\hat{x}, \hat{y})$ is a non-negative saddle point of $G(x, y)$,

then

2.a) $\hat{G}_x \leq 0$, $\hat{x}^i \hat{G}_{x^i} = 0$; $i = 1, \ldots, n$.

2.b) $G^o_{y^i} \geq 0$, $y^i_o G^o_{y^i} = 0$; $i = 1, \ldots, m$.

Proof: Let u_i be a vector with m components such that the ith component is 1 and all other components are zeros. Let v_i be a vector with m components all of which are zeros except the ith which is 1. Let $\Delta y_i = \frac{1}{h} [G(\hat{x} + hu_i, \hat{y}) - G(\hat{x}, \hat{y})]$, where h is a scaler. Similarly let $\Delta y_i = \frac{1}{h} [G(\hat{x}, \hat{y} + hv_i) - G(\hat{x}, \hat{y})]$, where h is a scaler.

Proof of 2.a: Let i_o be an arbitrary integer between 1 and n. Consider the following two cases:

Case 1: $\hat{x}^{i_o} > 0$. By 1.a, x^{i_o} maximizes the function

$$\omega(\hat{x}^i) = G(\hat{x} + hu_{i_o}, \hat{y}), \text{ for } x^{i_o} = \hat{x}^{i_o} + hu_{i_o} \geq 0.$$

Since $\hat{x}^{i_o} > 0$, h may be positive or negative for $x^{i_o} \neq \hat{x}^i$ and non-negative. Thus:

$$\Delta_{x^{i_o}} \leq 0 \text{ for } h > 0 \text{ and}$$

$$\Delta_{x^{i_o}} \geq 0 \text{ for } h < 0.$$

Passing to the limit as $h \to 0$, we get:

$$\hat{G}_{x^{i_o}} \leq 0 \text{ and } \hat{G}_{x^i} \geq 0.$$

showing, since $\hat{G}_{x^i}$ exists, that $\hat{G}_{x^{i_o}} = 0$. Thus $\hat{G}_{x^{i_o}} \leq 0$ (with the equality holding) and $\hat{G}_{x^{i_o}} \hat{x}^{i_o} = 0$.

Case 2: $x^{i_o} = 0$. Again, by 1.a, x^i maximizes $\omega(x^{i_o}) \geq 0$. However, $x^{i_o} = hu_{i_o}$, and $h > 0$ for $0 \leq x^{i_o} \neq \hat{x}^{i_o}$. Thus:

$$\Delta_{x^{i_o}} \leq 0 \text{ for } x^{i_o} \geq 0.$$

Now if $x^{i_o} < 0$, either $\omega(x^{i_o}) \geq \omega(\hat{x}^{i_o})$ or $\omega(x^{i_o}) \leq \omega(\hat{x}^{i_o})$. Since h must be negative we have:

$$\Delta_{x^{i_o}} \leq 0 \text{ or } \Delta_{x^{i_o}} \geq 0 \text{ for } h < 0.$$

Passing to the limit as $h \to 0$ we get:

$$\hat{G}_{x^{i_o}} \leq 0 \text{ or } \hat{G}_{x^{i_o}} = 0.$$

In either case $G_{x^{i_o}} \leq 0$.

This proves 2.a. The proof of 2.b is analogous and the lemma is proved.

Lemma 2). If

1) $G(x, y)$ is differentiable,

2) condition 2 of the conclusion of lemma 1 holds at $(\hat{x}, \hat{y})$ and if

3.a) $G(x, \hat{y}) \leq G(\hat{x}, \hat{y}) + \sum_{i=1}^{n} \hat{G}_{x^i} \xi^i$, $\xi^i = x^i - \hat{x}^i$, $x^i \geq 0$

3.b) $G(\hat{x}, y) \geq G(\hat{x}, \hat{y}) + \sum_{i=1}^{n} \hat{G}_{y^i} \eta^i$, $\eta^i = y^i - \hat{y}^i$, $y^i \geq 0$,

then $(\hat{x}, \hat{y})$ is a non-negative saddle point of $G(x, y)$.

Proof: We must show that 1.a and 1.b hold.

1.a)

i) $G(x, \hat{y}) \leq G(\hat{x}, \hat{y}) + \sum_{i=1}^{n} \hat{G}_{x^i} \xi^i = G(\hat{x}, \hat{y}) - \hat{G}_{x^i} \hat{x}_i + \sum \hat{G}_{x^i} x^i$ for all $x^i \geq 0$ by 3.a. But, by 2.a, $\hat{G}_{x^i} \hat{x}^i = 0$ and $\hat{G}_{x^i} \leq 0$. So that $\hat{G}_{x_i} x^i \leq 0$ for $x^i \geq 0$.

Thus

ii) $G(x, \hat{y}) - G(\hat{x}, \hat{y}) = \sum_{i=1}^{n} \hat{G}_{x^i} x^i \leq 0$ for all $x \geq 0$.

This shows that 1.a holds.

1.b)

By 3.b,

iii) $G(\hat{x}, y) \geq G(\hat{x}, \hat{y}) + \sum_{i=1}^{m} \hat{G}_{y^i}\eta^i = G(\hat{x}, \hat{y}) - \sum_{i=1}^{m} \hat{G}_{y^i}\hat{y}^i + \sum_{i=1}^{m} \hat{G}_{y^i} y^i$ for $y^i \geq 0$.

But, by 2.b, $\hat{G}_{y^i}\hat{y}^i = 0$ and $\hat{G}_{y_i} \geq 0$. Hence $\hat{G}_{y^i} y^i \geq 0$ for $y^i \geq 0$. Thus

iv) $G(\hat{x}, y) - G(\hat{x}, \hat{y}) = \sum_{i=1}^{m} \hat{G}_{y^i} \geq 0$, for all $y \geq 0$.

This shows that relation 1.b follows and the lemma is proved.

Next we provide sufficient conditions for 3.a and 3.b of lemma 2.

Lemma 3). If

1) $G(x, y)$ is differentiable,

2) $G(x, y)$ is concave in x,

3) $G(x, y)$ is convex in y,

then conditions 3.a and 3.b of lemma 2 are satisfied.

Proof: The proof is due to Kuhn and Tucker [32] and it involves the definition of convexity, concavity, and the differential. We again repeat that proof for completeness. By definition of concavity, if $x \neq \hat{x}$, we have

$$\theta G(x, \hat{y}) + (1 - \theta)G(\hat{x}, \hat{y}) \leq G[(1 - \theta)\hat{x} + \theta x, \hat{y}],\ 0 \leq \theta \leq 1,\ \text{i.e.}$$

$$\theta[G(x, \hat{y}) - G(\hat{x}, \hat{y})] \leq G[\hat{x} - \theta(x - \hat{x}), \hat{y}] - G(\hat{x}, \hat{y}).$$

Hence, for $0 < \theta \leq 1$

$$G(x, \hat{y}) - G(\hat{x}, \hat{y}) \quad \frac{1}{\theta}[G(\hat{x} - \theta(x - \hat{x}), \hat{y}) - G(\hat{x}, \hat{y})].$$

By the order preserving property of the limit and by definition of the differential - we have

$$G(x, \hat{y}) - G(\hat{x}, \hat{y}) \leq \sum_{i=1}^{n} (\hat{G}_{x^i})(x^i - \hat{x}^i) = \sum_{i=1}^{n} \hat{G}_{x^i}\xi^i,$$

which proves that condition 3.a of lemma 2 holds. Analogously we could show that condition 3.b of lemma 2 holds by using the definition of convexity and of the differential. The proof of the lemma is complete.

Lemma 4). Let A be an open convex subset of E^n. And let $f(x)$ be twice of class C^2 on A. $f(x)$ is concave on A if and only if

$$Q(x, t) = \sum_{i=1}^{n}\sum_{j=1}^{n} f_{ij}(x)t_i t_j$$

be less than or equal to zero for all t and for all x in A.

Proof: The proof uses the definition of concavity and the mean value theorem.

Let x_1 and x_2 be two points in A and let $\bar{x} = \theta x_1 + (1 - \theta)x_2$ for $0 < \theta < 1$.

Sufficiency:

By definition of concavity we must show that

i) $\theta f(x_1) + (1 - \theta)f(x_2) - f(\bar{x}) \leq 0$.

But the left hand side of (i) may be written as

$$\theta[f(x_1) - f(\bar{x})] + (1 - \theta)[f(x_2) - f(\bar{x})]$$

By the second order mean value theorem there exists a point x' on the line segment connecting x_1 and $\bar{x}$ such that

$$f(x_1) - f(\bar{x}) = \sum_i f_i(\bar{x})(x_1^i - \bar{x}^i) + 1/2 \sum_i \sum_j f_{ij}(x')(x_1^i - \bar{x}^i)(x_1^j - \bar{x}^j) =$$

$$= \sum_i f_i(\bar{x})(1 - \theta)(x_1^i - x_2^i) + \sum_{i,j} f_{ij}(x')(1 - \theta)^2(x_1^i - x_2^i).$$

So that

(ii) $$\theta[f(x_1) - f(\bar{x})] = \theta[\sum_i f_i(\bar{x})(1 - \theta)(x_1^i - x_2^i) + $$

$$+ 1/2 \sum_{i,j} f_{ij}(x')(1 - \theta)^2(x_1^i - x_2^i)(x_1^j - x_2^i)].$$

Similarly, there exists a point x" on the line segment connecting x_2 and $\bar{x}$ such that

(iii) $$(1 - \theta)[f(x_2) - f(\bar{x})] = (1 - \theta)[\sum_i f_i(\bar{x})\theta(x_2^i - x_1^i) +$$

$$+ 1/2 \sum_{i,j} f_{ij}(x'')\theta^2(x_2^i - x_1^i)(x_2^j - x_1^i)].$$

Adding (ii) and (iii) we get

(iv) $$\theta[f(x_1) - f(\bar{x})] + (1 - \theta)[f(x_2) - f(\bar{x})] = \theta(1 - \theta)\{\sum_i f_i(\bar{x})[(x_1^i - x_2^i)$$

$$+ (x_2^i - x_1^i)] + \theta Q_1 + (1 - \theta)Q_2\} = \theta Q_1 + (1 - \theta)Q_2.$$

where Q_1 and Q_2 are the quadratic forms in (ii) and (iii) respectively. Now x' and x" are in A by convexity of A, and hence Q_1 and Q_2 are non-positive. So that $\theta f(x_1) + (1 - \theta)f(x_2) - f(\bar{x}) \leq 0$. And the sufficiency part of the lemma is proved.

Necessity:

Suppose, by way of contradiction that x_o is a point in A such that $Q(x, t) = Q(x_o, t_o)$ is positive for some t. Then, by continuity of Q, there exists a neighborhood $N(x_o)$ of x_o such that $Q(x, t_o)$ is positive for $x \in N(x_o)$. Let $x_2 \in N(x_o)$ be such that $x_2 - x_o = \lambda t$ for $\lambda \neq 0$. Then we have, by repeating the calculations that lead to (iv),

$$\text{(v)}\quad \theta f(x_o) + (1-\theta)f(x_2) - f(\bar{x}) = 1/2\ \theta(1-\theta)^2[\sum_{i,j} f_{ij}(x')\lambda^2 t^i t^j] +$$

$$+ 1/2\ (1-\theta)\theta^2[\sum_{i,j} f_{ij}(x'')\lambda^2 t^i t^j], \text{ where } \bar{x} = \theta x_o + (1-\theta)x_2,$$

$0 < \theta < 1$, x' is on the line segment connecting $\bar{x}$ and x_o and x'' is on the line segment connecting $\bar{x}$ and x_2. Noting that the left hand side of (v) is positive we obtain a contradiction to the concavity of $f(x)$. This proves lemma 4.

We now have the necessary equipment to develop the theorems of this section.

The problem we are concerned with here, is the maximization of a function $f(x)$, $x \in E^n$, subject to $h^\beta(x) \geq 0$, $\beta = 1, \ldots, \ell$ and to $x \geq 0$. Thus a local constrained maximum at x_o means that there exists a neighborhood of x_o such that $f(x_o) \geq f(x)$ for all $x \geq 0$ in that neighborhood satisfying $h^\beta(x) \geq 0$, $\beta = 1, \ldots, \ell$.

Theorem 1. If 1) $f(x_o)$ is a local constrained maximum, 2) the function f and h^β have continuous second order derivatives, 3) the rank of Jacobian of effective constraints h^β, at $\hat{x}$, evaluated at $\hat{x}$ is equal to the number of effective constraints, 4) the number of effective constraints plus the number of zero components of $\hat{x}$ is less than n, 5) f and h^β are concave in x, then there exists a vector $\hat{\mu} = (\hat{\mu}^1, \ldots, \hat{\mu}^\ell)$ with $\hat{\mu}^\beta \geq 0$, $\beta = 1, \ldots, \ell$, such that $L(\hat{x}, \hat{\mu})$ is a non-negative saddle value of $L(x, \mu) = f(x) + \Sigma\mu^\beta h^\beta$.

Proof: The proof of the theorem consists of verifying the conditions of lemma 1 of this section. This, in turn, is done by showing that we may apply theorem 1 of Chapter 3 to our present problem.

Let $h^{\ell+i}(x) = x^i$, $i = 1, \ldots, n$. Then by hypothesis $f(x_o)$ is a local maximum of $f(x)$ subject $h^\beta(x) \geq 0$ and $h^{\ell+i}(x) \geq 0$, $\beta = 1, \ldots, \ell$; $i = 1, \ldots, n$. Let $h^{(1)}(x)$ be the vector whose components are the h^β's and let $h^{(2)}(x)$ be the vector whose components are the h^{m+i}'s. Let $h^{\Gamma_i}(x)$ be the vector whose components are the zero components of $h^{(i)}(\hat{x})$, $i = 1, 2$. Let $h^\Gamma(x) = (h^{\Gamma_1}, h^{\Gamma_2})$. The matrix $h_x^\Gamma(x)$ has the form

$$\begin{bmatrix} h^{\Gamma_1}(\hat{x}) \\ h^{\Gamma_2}(\hat{x}) \end{bmatrix}$$

where the component $h_j^i(\hat{x})$ of $h_x^{\Gamma_2}(\hat{x})$ is 1 when $i = j$ and zero otherwise. One may, by remumbering the variables, choose a square submatrix of $h_x^\Gamma(\hat{x})$ of order equal to the number of rows of that matrix that is non-singular. Thus the rank condition of theorem 1 of Chapter 3 is satisfied. The other conditions of that theorem are satisfied by virtue of the hypothesis of the present theorem. Thus there exists a vector of multipliers $\bar{\mu} = (\bar{\mu}^1, \bar{\mu}^2) \geq 0$ such that

(i) $L_{x^i}(\hat{x}, \bar{\mu}) = 0$, where $L(\hat{x}, \bar{\mu}) = f(x) + \sum_{\beta=1}^{\ell} \bar{\mu}^{\beta} h^{\beta}(x) + \sum_{i=1}^{n} \bar{\mu}^{\ell+i} h^{\ell+i}(x)$.

(ii-a) $\bar{\mu}^{\beta} h^{\beta}(\hat{x}) = 0, \ \beta = 1, \ldots, \ell$

(ii-b) $\bar{\mu}^{\ell+i} h^{\ell+i}(\hat{x}) = 0, \ i = 1, \ldots, n.$

Now

$$L_{x^i}(\hat{x}, \bar{\mu}) = f_i(\hat{x}) + \sum_{\beta=1}^{\ell} \bar{\mu}^{\beta} h_i^{\beta}(\hat{x}) + \bar{\mu}^{\ell+i} = 0.$$

Take $\hat{\mu} = \bar{\mu}^1$. Then $L_{x^i}(\hat{x}, \bar{\mu}) = L_i(\hat{x}, \hat{\mu}) + \bar{\mu}^{\ell+i} = 0$. But $\bar{\mu}^{\ell+i} \leq 0$ and so

(iii) $L_{x^i}(\hat{x}, \hat{\mu}) \leq 0$.

Also, by (ii-b), the equality holds if $\hat{x}^i \geq 0$, i.e.

(iv) $\hat{L}_{x^i} \leq 0, \ \hat{x}^i \hat{L}_{x^i} = 0, \ i = 1, \ldots, n.$

Noting that $\hat{x} \geq 0$ and $h^{\beta}(\hat{x}) \geq 0$, by hypothesis of this theorem, we have

(v) $L_{\mu^{\beta}}(\hat{x}, \hat{\mu}) = h^{\beta}(\hat{x}) \geq 0, \ \beta = 1, \ldots \ell.$

Furthermore, by (ii-a),

(vi) $\hat{\mu}^{\beta} h^{\beta}(\hat{x}) = 0.$

By (v) and (vi) we have

(vii) $\hat{L}_{\mu^{\beta}} \geq 0, \ \hat{\mu}^{\beta} h^{\beta}(\hat{x}) = 0, \ \beta = 1, \ldots, \ell.$

We may now apply lemma 2 above to the function $L(x, \mu)$. By (iv), (vii), the fact that $(\hat{x}, \hat{\mu}) \geq 0$, assumption 5 of this theorem, concavity of $L(x, \mu)$ in x and linearity--hence convexity--of $L(x, \mu)$ in μ the lemma applies. QED.

Theorem 2. _If there exists an m-dimensional non-negative vector μ_o such that $L(x_o, \mu_o)$ is a non-negative saddle value of $L(x, \mu)$, where $L(x, \mu)$ is as in theorem 1, then $f(\hat{x})$ is a constrained maximum in the sense of this section._

Proof: The proof consists of simply applying the definition of a non-negative saddle point to the function $L(x, \mu)$. Doing so, we have

$$L(x, \hat{\mu}) \leq L(\hat{x}, \hat{\mu}), \ x \geq 0, \text{ i.e.}$$

(i) $f(x) + \sum_{\beta} \hat{\mu}^{\beta} h^{\beta}(x) \leq f(\hat{x}) + \sum \hat{\mu}^{\beta} h^{\beta}(\hat{x}), \ x \geq 0.$

Since $L(\hat{x}, \hat{\mu})$ is a saddle value for $L(x, \mu)$ we have, by lemma 1,

(ii) $\hat{\mu}^{\beta} h^{\beta}(\hat{x}) = 0.$

Thus, by (i), $f(x) + \sum_{\beta} \hat{\mu}^{\beta} h^{\beta}(x) \leq f(\hat{x})$, i.e.

(iii) $f(\hat{x}) - f(x) \geq \sum_{\beta} \hat{\mu}^{\beta} h^{\beta}(x), \ x \geq 0.$

Noting that the right hand side of the inequality (iii) is non-negative for any x satisfying $h^{\beta}(x) \geq 0$, $\beta = 1, \ldots, \ell$, we have $f(\hat{x}) \geq f(x)$ for any $x \geq 0$ satisfying $h^{\beta}(x) \geq 0$, $\beta = 1, \ldots, \ell$. QED.

Corollary 1) If conditions 2, 3, 4 and 5 of theorem 1 are satisfied then $f(\hat{x})$ is a local constrained maximum if and only if there exists $\hat{\mu} \geq 0$ such that $L(\hat{x}, \hat{\mu})$ is a non-negative saddle value for $L(x, \mu)$.

Proof. The sufficiency part follows directly from theorem 2 and the necessity part follows, also directly, from theorem 1.

Remark 1) Let us define a local non-negative saddle point as a non-negative point that satisfies inequalities 1.a) for $x \geq 0$ in a neighborhood of $\hat{x}$ and similarly with 1.b. Theorems 1 and 2 and Corollary 1 hold if we replace "non-negative saddle point" by "local non-negative saddle point".

Remark 2) Corollary 1 is quite expensive for, as clear from theorem 2, all that is needed for the sufficiency part is that $L(\hat{x}, \hat{\mu})$ be a non-negative saddle point.

Theorem 3. If 1) $f(x)$ and $h^{\beta}(x)$, $\beta = 1, \ldots, \ell$, have continuous second order derivatives, 2) there exists a vector $\mu \geq 0$ such that $\hat{L}_{x_i} = 0$, $i = 1, \ldots, n$, where $\hat{L}(x, \bar{\mu})$ is as defined in the proof of theorem 1, 3) $Q(\hat{x}, \eta) = \eta \hat{L}_{xx} \eta^* < 0$ for $\eta \neq 0$. 4) $h^{\beta}(\hat{x}) \geq 0$, $\beta = 1, \ldots, \ell$, then there exists a neighborhood of $\hat{x}$, such that $f(\hat{x})$ is a local constrained maximum.

First proof of theorem 3: Clearly condition 3) implies that condition 4) of theorem 4 of Chapter 3 is satisfied. Furthermore, all other conditions of that theorem hold and the present theorem follows.

Second proof: By an arguement similar to that used in the proof of theorem 1 we have relations (iv) and (vii) in that proof. By assumption 3) of this theorem, note that by continuity of the elements of L_{xx} there exists an open sphere with center at $\hat{x}$ such that $\eta F_{xx}\eta^* < 0$ for x in that sphere. Thus by lemma 4, $L(x, \mu)$ is concave in x in that sphere.

Finally, by linearity in μ, $L(x, \mu)$ is convex in μ. Thus lemma 2 applies, locally, and $L(\hat{x}, \hat{\mu})$ is a local non-negative saddle point of $L(x, \mu)$. Now the conclusion follows from theorem 2 if we confine our attention to the sphere mentioned above.

Remark 3) The two ways of proving theorem 3 indicate the relation between the two types of sufficiency theorems but do not give their exact relationship. They show that sufficient conditions for a local non-negative saddle point imply the sufficient conditions for a local constrained maximum.

2. Extension to Vector Maxima.

Let $F(x)$ be a k-dimensional vector valued function of x, $F(x) = (f^1(x), \ldots, f^k(x))$.

Definition 1) $F(\hat{x})$ is said to be a constrained vector maximum of $F(x)$---subject to the constraint $h^\beta(\bar{x}) \geq 0$, $\beta = 1, \ldots, \ell$, $x \geq 0$--if there does not exist any $\bar{x} \geq 0$ with $h^\beta(\bar{x}) \geq 0$, $\beta = 1, \ldots, \ell$, such that $f^j(\bar{x}) \geq f^j(\hat{x})$ with strict inequality for at least one j.

Remark: It follows from the above definition that if $F(\hat{x})$ is a constrained vector maximum, then $f^{j_o}(\hat{x})$ is a maximum subject to: $f^j(x) \geq f^j(\hat{x})$, $j \neq j_o$, between 1 and k, and to: $h^\beta(x) \geq 0$, $\beta = 1, \ldots, \ell$; $x \geq 0$, for $j_o = 1, \ldots, n$. For if not, then there exists an $\bar{x} \geq 0$ and a $\bar{j}$ such that:

$$f^{\bar{j}}(\bar{x}) \geq f^{\bar{j}}(\hat{x})$$

$$f^{j}(\bar{x}) \geq f^{j}(\hat{x}),\ i \neq \bar{i} \text{ between 1 and n},$$

$$h^\beta(\bar{x}) \geq 0.$$

Utilizing this remark we provide a calculus proof of a theorem that is closely related to theorem 4 of Kuhn and Tucker [32]. The method of the proof was suggested by Professor L. Hurwicz.

Definition 2) Fix j_o. A point $\hat{x}$ is said to be j_o-regular of the matrix

$$\begin{bmatrix} \hat{F}^{\Gamma}_x \\ \hat{h}^{\Gamma}_x \end{bmatrix}$$

has maximum rank, where $F^{\Gamma}(x)$ is a vector whose components are f^j's with $j \neq j_o$, $h^{\Gamma}(x)$ is the vector whose components are $h^{\beta}(x)$ such that $h^{\beta}(\hat{x}) = 0$.

Definition 3) A point $\hat{x}$ is said to be regular if it is j_o-regular for $j_o = 1, \ldots, n$.

Definition 4) $\hat{x}$ is said to be j_o-constrained maximum if statement (I) holds.

Theorem 4. If 1) $F(\hat{x})$ is a constrained vector maximum, 2) $\hat{x}$ is regular, 3) $f^j(x)$ and h (x) have continuous second order partial derivatives, 4) ℓ + the number of effective constraints, including $x^i \geq 0$, is less than n, then there exist constants $\hat{\nu}^j > 0$, $j = 1, \ldots n$, and a m-vector $\hat{\mu} \geq 0$ such that

(i) $\hat{L}_{x^i} \leq 0$, $\hat{x}^i\hat{L}_{x^i} = 0$, $i = 1, \ldots, n$

(ii) $\hat{L}_{\mu^\beta} \geq 0$, $\hat{\mu}^\beta\hat{L}_{\mu^\beta} = 0$, $\beta = 1, \ldots, \ell$,

where $L(\mu, x) = \Psi(x) + \sum_{\beta=1}^{\ell} \mu^\beta h^\beta(x)$, and $\Psi(x) = \sum_{j=1}^{\ell} \hat{\nu}^j f^j(x)$.

Proof: By the remark at the beginning of this section, $\hat{x} j_o$-constrained maximum for $j_o = 1, \ldots, n$. Hence $\hat{x}$ is a locally j_o-constrained maximum, i.e. $f^{j_o}(\hat{x})$ is a local maximum subject to:

$$h^\beta(x) \geq 0$$
$$h^{m+j}(x) = f^j(\hat{x}) \geq 0, \ j \neq j_o,$$
$$h^{m+(\ell-1)+i}(x) = x^i \geq 0.$$

Fix j_o. Since x is a j_o-regular, the rank condition in theorem 1 of Chapter 2 is satisfied, by an argument similar to that of the proof of theorem 1 of this chapter. We get, again using the same argument we used in proving relations (iv) and (vii) in the proof of theorem 1,

there exist vectors $\lambda^{j_o} \geq 0$ and $\mu^{j_o} \geq 0$, where the j_oth component of the ℓ-dimensional vector $\lambda_o^{j_o}$ is equal to one, such that

(i) $\hat{L}^{j_o}_{x^i} \leq 0$, $\quad x_o^i\hat{L}^{j_o}_{x_i} = 0$, $\quad i = 1, \ldots, n$,

(ii) $\hat{L}^{j_o}_{\lambda^{j_o}} \geq 0$, $\quad \lambda_o^{j_o}\hat{L}^{j_o}_{j_o, j} = 0$ $\quad j \neq j_o$,

(iii) $L^{j_o}_{\mu^{j_o,\beta}} \geq 0, \qquad \mu_o^{j_o,\beta} L^{j_o}_{j_o,\beta} = 0, \qquad \beta = 1, \ldots, \ell,$

where

$$L^{j_o}(\lambda^{j_o}, \mu^{j_o}, x) = f^{j_o}(x) + \sum_{j \neq j_o} \lambda^{j_o, j}[f^j(x) - f^j(x_o)] +$$

$$+ \sum_{\beta} \mu^{j_o, \beta} h^{\beta}(x).$$

Rep ating the above argument as j_o goes from 1 to k we have for $j_o = 1, \ldots, \ell$

(iv) $\hat{L}^{j_o}_{x_i} = \hat{f}^{j_o}_{x_i} + \sum_{j \neq j_o} \lambda^{j_o, j}\hat{f}^j_{x_i} + \sum_{\beta} \mu^{j_o, \beta}\hat{h}^{\beta}_{x^i} \leq 0, \; \hat{x}^i \hat{L}^{j_o}_{x^i} = 0, \; i = 1, \ldots, n.$

(v) $\hat{L}^{j_o}_{\lambda^{j_o}, j} = f^j(\hat{x}) - f^j(\hat{x}) = 0, \; j \neq j_o$

(vi) $\hat{L}^{j_o}_{\mu^{j_o}, \beta} = h^{\beta}(x) \geq 0, \; \mu^{j_o, \beta}\hat{L}^{j_o}_{\mu_o^{j_o}, \beta} = 0, \; \beta = 1, \ldots, m.$

Summing each of systems (iv) and (vi) over j_o we get:

(vii) $\sum_{j=1}^{\ell}(1 + \sum_{j_o \neq j} \lambda^{j_o, j})\hat{f}^j_{x^i} + \sum_{\beta=1}^{m}(\sum_{j_o=1} \mu^{j_o, \beta})\hat{h}^{\beta}_{x^i} \leq 0$, with equality if

$\hat{x}^i > 0, \; i = 1, \ldots, n.$

(viii) $\ell h^{\beta}(\hat{x}) \geq 0, \; (\sum_{j_o=j}^{\ell} \mu^{j_o, \beta})h(\hat{x}) = 0, \; \beta = 1, \ldots, \ell$, i.e.

(ix) $h^{\beta}(\hat{x}) \geq 0, \; (\sum_{j_o=1}^{\ell} \mu^{j_o, \beta})h^{\beta}(\hat{x}) = 0.$

Now let $\hat{\nu}^j = 1 + \sum_{j_o \neq j} \lambda_o^{j_o, j}$, and let $\hat{\mu}^{\beta} = \sum_{j_o=1}^{\ell} \mu^{j_o, \beta}$. Clearly $\hat{\nu}^j > 0$ for $j = 1, \ldots, \ell$, and $\hat{\mu}^{\beta} \geq 0$ for $\beta = 1, \ldots m$. Consider

$$L(x, \mu) = g(x) + \sum_{\beta=1}^{\ell} \mu^{\beta}h^{\beta}(x), \text{ where } g(x) = \sum_{j=1}^{\ell} \hat{\nu}^j f^j(x).$$

The conclusion of the theorem follows from (vii) and (ix) above.

Corollary 1) If, in addition to the hypothesis of theorem 4, the functions $f^j(x)$ and $h^{\beta}(x)$ are concave, $j = 1, \ldots, \ell$; $\beta = 1, \ldots, m$, then $(\hat{x}, \hat{\mu})$ is a non-negative saddle point for $\Psi(x, \mu)$.

Proof: The corollary follows by combining the results of theorem 4 and lemma 2 of Section 1 taking into consideration that $L(x, \mu)$ is linear, hence convex, in μ.

3. A discrete time optimal control problem.

Consider a control system where the equations of motion are given by the system of difference equations:

(1) $x^{t+1} - x^t = f^t(x^t, u^t)$, $t = 1, \ldots, T - 1$,

where x^t is an n-vector and where u^t is an m-vector. Let x denote the finite sequence $\{x_1, \ldots, x_T\}$ and let u denote the finite sequence $\{u_1, \ldots, u_T\}$. We shall consider two problems:

Problem I: Maximize $J(x, u)$ subject to (1) and to

(2.1) $\bar{h}^t(x^t, u^t) = 0$, $1 \leq t \leq T - 1$

(2.2) $\bar{\bar{h}}^t(x^t, u^t) \geq 0$, $1 \leq t \leq T - 1$, where $\bar{h}$ is an N_1-vector function and $\bar{\bar{h}}$ is an N_2-vector function.

(3.1) $\bar{g}(x^1, x^T) = 0$,

(3.2) $\bar{\bar{g}}(x^1, x^T) \quad 0$, where $\bar{g}$ is an M_1-vector function and $\bar{\bar{g}}$ is an M_2-vector function.

Problem II: Maximize $\sum_{t=1}^{T-1} R^t(x^t, u^t) + R^T(x^T)$ subject to (1), (2) and (3).

Problem II is a special case of problem I, and is the more familiar form of the optimal control problem. We shall only provide a brief introduction to the problem. The reader is referred to Canon, Cullum and Polak [49] for more detailed and general treatment of the problem and for references.

3.1 Characterization of solutions of Problem I.

Let $\mathring{z} = (\mathring{x}, \mathring{u})$ be a given vector in the $nT + m(T - 1)$-dimensional Euclidian space.

Proposition 1: If: 1) J, f, h and g are of class C^1. 2) z is a solution to problem I. Then there exist vectors $(\lambda_o, \bar{\gamma}, \bar{\bar{\gamma}})$ and sequences $\{\lambda^t, \bar{\mu}^t, \bar{\bar{\mu}}^t\}$ with $(\lambda_o, \gamma, \lambda, \mu) \neq 0$, $\lambda_o \geq 0$, such that:

(4.1) $\bar{\bar{\gamma}} \geq 0$, $\bar{\bar{\gamma}}\bar{\bar{g}}(\hat{z}) = 0$

(4.2) $\bar{\bar{\mu}}^t \geq 0$, $\bar{\bar{\mu}}^t\bar{\bar{h}}^t(\hat{z}) = 0$, $t = 1, \ldots, T - 1$.

(5) $\lambda^{t-1} - \lambda^t = \lambda_o \hat{J}_{x^t} + \lambda^t \hat{f}^t_{x^t} + \mu^t \hat{h}^t_{x^t}$, $t = 2, \ldots, T - 1$.

(6) $\lambda_o \hat{J}_{u^t} + \lambda^t \hat{f}^t_{u^t} + \mu^t \hat{h}^t_{u^t} = 0$, $t = 1, \ldots, T - 1$.

(7.1) $\lambda_o \hat{J}_{x^1} + \lambda^1 + \lambda^1 \hat{f}^1_{x^1} + \mu^1 \hat{h}^1_{x^1} + \gamma \hat{g}_{x^1} = 0.$

(7.2) $\lambda_o \hat{J}_{x^T} + \gamma \hat{g}_{x^T} = 0.$

Proposition 2. If 1) J, f, h and g are of class C^1. 2) f, $\bar{h}$, and $\bar{g}$ are linear. 3) J, $\bar{\bar{h}}$, $\bar{\bar{g}}$ are concave. 4) Conditions (4)-(6) of proposition 1 hold at a point z which satisfies constraints (1)-(3). Then z is a solution to problem 1.

Outline of proofs: To prove proposition 1, apply theorem 1 of chapter 1. Define $F^I = J + \sum_{t=1}^{T-1} \lambda^t(f^t(x^t, u^t) - x^{t+1} + x^t + \sum_{t=1}^{T-1} \mu^t h^t + \gamma g$. The existence of the multipliers λ_o, λ, μ, γ is simplified by the theorem and they satisfy properties (4) of proposition 1. Writing down the fact that the derivatives of F with respect to x^t and u^t and equating the expressions at z to zero we have:

(8) $\hat{F}^I_{u^t} = \lambda_o \hat{J}_{u^t} + \lambda^t \hat{f}^t_{u^t} + \mu^t \hat{h}^t_{u^t} = 0,\ t = 1, \ldots, T - 1.$

(9) $\hat{F}^I_{x^1} = \lambda_o \hat{J}_{x^1} + \lambda^1 + \lambda^1 \hat{f}^1_{x^1} + \mu^1 \hat{h}_{x^1} + \gamma \hat{g}_{x^1} = 0$

(10) $\hat{F}^I_{x^t} = \lambda_o \hat{J}_{x^t} + \lambda^t - \lambda^{t-1} + \lambda^t \hat{f}^t_{x^t} + \mu^t h^t_{x^t} = 0,\ t = 1, \ldots, T - 1.$

(11) $\hat{F}^I_{x^T} = \lambda_o \hat{J}_{x^T} + \hat{g}_{x^T} = 0.$

Conditions (10), (8), (9) and (11) are equivalent to conditions (5), (6), (7.1) and (7.2), respectively, of proposition 1.

To prove proposition 2, we observe that theorem 3 of chapter 1 applies, in view of (8)-(11), and in view of assumptions (2) and (3) of the proposition.

3.2 Characterization of problem II.

Proposition 3. If 1) R, f, h and g are of class C^1. 2) $\hat{z}$ is a solution to problem II. Then there exists a vector (λ_o, γ) and sequences $\{\lambda^t\}$ and $\{\mu^t\}$ with $(\lambda_o, \gamma, \lambda^t, \mu^t) \neq 0$ for t = 1, ..., T - 1, such that, in addition to (4) in proposition 1, we have:

(12) $\lambda^{t-1} - \lambda^t = \hat{H}^t_{x^t}$, t = 2, ..., T - 1, where $H^t = \lambda_o R^t + \lambda^t f^t + \mu^t h^t$,

(13) $\hat{H}_{u^t} = 0$, t = 1, ..., T.

(14.1) $\hat{G}_{x^1} + \hat{H}^1_{x^1} + \lambda^1 = 0$, where $G = \lambda_o R^1 + \lambda_o R^T + \gamma g$.

(14.2) $\hat{G}_{x^T} = 0$.

Proposition 4) If 1) R, f, h and g are of c ss C^1. 2) $\hat{z}$ satisfies constraints (1)-(3). 3) f, $\bar{h}$ and $\bar{g}$ are linear. 4) R, $\bar{\bar{h}}$, $\bar{\bar{g}}$ are concave. 5) Conditions (4), (12)-(14) are satisfied at $\hat{z}$. Then $\hat{z}$ is a solution to problem II.

Propositions 3) and 4) follow immediately from propositions 1) and 2). Compare propositions 3) and 4) to the results in section 1 of chapter 7.

4. Some Economic Applications.

In this section we apply some of the theorems of Chapter 3 and of the present chapter derive some of the classical propositions of economic theory.

4.1 Consumer's Optimum.

Let $x = (x_1, \ldots, x_n)$ denote a commodity bundle. We speak of a commodity space as the set of all commodity bundles and denote it by $\mathfrak{X}$. We define a consumer as an entity with: a) A complete ordering of the commodity space and b) An income, denoted by m. Let $p = (p_1, \ldots, p_n)$ denote a price vector. The expenditure of the consumer, if he buys a bundle x, is p.x, where $p.x = \sum_{i=1}^{n} p_i n_i$. We postulate that the consumer chooses the bundle which is most preferred subject to the budget constraint $p.x \leq m$ and provided that x lies in a subset of $\mathfrak{X}$ which we call the consumption set and denote by $\mathfrak{X}^*$. We shall assume, further, that:

A.1) $\mathfrak{X}$ is the Euclidian n-space.

A.2) The consumer's ordering is representable by a real valued function U(x).

A.3) That $\mathfrak{X}^*$ is the non-negative orthant of $\mathfrak{X}$, i.e. $\mathfrak{X}^* = \{x \in \mathfrak{X} : x \geq 0\}$.

We now present some of the well known propositions about consumer's optimum. We start by introducing some lemmas which indicate the characterization of a consumer's optimum.

Lemma 1: (first order necessary conditions) If

1) The function U is continuously differentiable,

2) The consumer takes prices parametrically, i.e. p is independent from x,

3) $\hat{x}$ is a consumer's optimum (i.e. $\hat{x}$ maximizes U(x) subject to $m - p.x \geq 0$ and $x \geq 0$),

4) Not all prices are zero (i.e. $p \neq 0$).

Then there exist a positive constant λ_o and a non-negative constant μ such that:

1) $\lambda_o \hat{U}_i \leq \mu p^i$ with equality if $\hat{x}_i > 0$, $(i = 1, \ldots, n)$,

2) $\mu(m - p.\hat{x}) = 0$, where $\hat{U}_i = \left.\frac{\partial U}{\partial x^i}\right|_{x = \hat{x}}$ and p^i is the ith component of p.

Proof: $\hat{x}$ is a solution to the following forms: max U(x) subject to $h^1(x) = m - p.x \geq 0$, $h^{1+i}(x) = x^i \geq 0$, $i = 1, \ldots, n$. Theorem 1 of section 1 applies (if the budget constraint is effective at $\hat{x}$, then condition 4 of the lemma guarantees that the rank condition is satisfied). Thus there exist a positive constant λ_o and non-negative constants $\mu, \gamma^1, \ldots, \gamma^n$ such that: a) $\hat{F}_i = 0 (i = 1, \ldots, n)$, where $F = \lambda_o U(x) + \mu(m - p.x) + \sum_{i=1}^{n} \gamma^i x^i$ and $\hat{F}_i = \frac{\partial F}{\partial x^i} \Big|_{x = \hat{x}}$, b) $\mu(m - p.\hat{x}) = 0$, $\gamma^i \hat{x}^i = 0$. From a) and b) it follows that : c) $\lambda_o \hat{U}_i - \mu p^i + \gamma^i = 0$, $\gamma^i \hat{x}^i = 0$, and d) $\mu(m - p.x) = 0$. d) establishes conclusion 2) of the lemma. From c) we have $\lambda_o U_i - \mu p^i = -\gamma^i$, $\gamma^i x^i = 0$. But $-\gamma^i \leq 0$ and conclusion 1) of the theorem is proved.

Lemma 2: (second order necessary conditions) If assumption 1) of lemma 1 is replaced by 1)' the function U has continuous second order derivatives. And if assumptions 2) and 4) of lemma 1 hold, then there exist λ_o and μ as in lemma 1 such that $\sum_{i,j} \hat{U}_{ij} \eta^i \eta^j \leq 0$ for all $\eta \neq 0$ satisfying,

(i) $p.\eta = 0$ whenever $m = p.\hat{x}$

(ii) $\eta^i = 0$ whenever $\hat{x}^i = 0$,

where $U_{ij} = \frac{\partial^2 U}{\partial x^i \partial x^j} \Big|_{x = \hat{x}}$ and where η^i and η^j denote components of η.

Proof: Theorem 3 of Chapter 3 applies. Thus, with F as in the proof of lemma 1, $F_{ij} = U_{ij}$. The conditions on η (given by (i) and (ii)) follow by differentiating h^1 and h^{1+i} with respect to x (where h^1, h^{1+i} are as in the proof of lemma 1). We then have $\lambda_o \sum_{i,j} U_{ij} \eta^i \eta^j \leq 0$ for $\eta \neq 0$ satisfying (i) and (ii). Since $\lambda > 0$ our conclusion follows.

We define a demand function as a function $H : E_+^{n+1} \to E_+^n$, where E_+^{n+1} is the non-negative orthant of the Euclidian (n + 1)-space whose elements are prices and income (p, m) and where E_+^n is the non-negative orthant of the Euclidian n-space whose elements are commodity bundles x. We write demand functions as $x^i = H^i(p, m)$. Suppose H^i to be differentiable. Then we define a Slutsky term as a change in the demand for commodity i due to a "compensated" change in the price of commodity j. We take compensation to mean a change in m such that the consumer could afford his previous optimal bundle. Formally, denoting Slutsky terms by K_{ij}, we write:

$$(3) \quad K_{ij} = \frac{\partial H^i}{\partial p_j} \Big|_{p.dx = 0} = H_j^i + x^j H^i{}_m$$

Proposition 1) If 1) The function U has continuous second order derivatives. 2) The demand functions are solutions to the consumer's optimum problems with $p.x = m$ and $x^i > 0$. 3) H^i are continuously differentiable. 4) Not all prices are zero. Then the demand functions have the following properties:

(i) H^i are homogeneous of degree zero in prices and income,

(ii) H^i satisfy $p.H = m$,

(iii) The Slutsky form $\sum_{i,j} K_{ij} dp^i dp^j \leq 0$.

Proof: Since the demand functions are obtained as solutions to the consumer's optimum problem, by lemma 1 and assumption 3, we have:

(4) $U_i = \bar{\mu} p_i$, where $\bar{\mu} = \mu/\lambda_o$

(5) $p.x = m$.

Properties (i) and (ii) follow from (5) which must be satisfied at all p and m.

To show that (iii) holds we first write:

(6) $dx^i = dH^i = \sum_j H^i_j dp^j + H^i_m dm.$

Differentiating (5) we have:

(7) $\sum_j p^j dx^j + \sum_i x^j dp^j = dm.$

Taking compensation, $\sum_j p^j dx^j = 0$, into consideration we have:

(7)' $dm = \sum_j x^j dp^j.$

Substituting from (7)' in (6) we have:

(8) $dx^i = \sum_j H^i_j dp^j + H^i_m \sum_j x^j dp^j.$

Multiplying each of the equations (8) by dp^i and summing we get:

(9) $\sum_i dx^i dp^i = \sum_{i,j} (H^i_j + H^i_m x^j) dp_i dp_j = \sum_{i,j} dp^i dp^j.$

Thus, property (iii) is demonstrated once we show that $\sum_i dx^i dp^i \leq 0$ for $p^i dx^i = 0$.

Differentiating (4) we have:

(10) $\sum_j U_{ij} dx^j = \bar{\mu} dp^i + p^i d\mu.$

Multiplying equations (10) by dx^i and adding the resulting equations we have:

(11) $\sum_{i,j} U_{ij} dx^i dx^j = \bar{\mu} \sum_i dp^i dx^i + d\mu \sum p^i dx^i.$

Taking note of the compensation, the last term on the right hand of (11) is zero and

(12) $\bar{\mu} \sum_i dx^i dp^i = \sum_{i,j} U_{ij} dx^i dx^j.$

By lemma 2, since $\sum_i p^i dx^i = 0$, the right hand side of (12) is non-positive. Since $\bar{\mu} \geq 0$ we have: $\sum_i dx^i dp^i \leq 0$, and property (iii) holds.

Remark 1: If we assume that the matrix $[U_{ij}]$ has an inverse, then it follows from (8), (10) and the symmetry of $[U_{ij}]$ (which follows from assumption 1) of proposition 1) that $[K_{ij}]$ is symmetric.

Remark 2: The proposition is true if we replace the utility index U(x) by any monotone increasing transformation of U.

Remark 3: If we define the marginal rate of transformation between x^i and x^j MRS_{ij}, $\frac{dx^i}{dx^j}$ along a given indifference curve U(x) = constant, then $MRS_{ij} = -\frac{U_i}{U_j}$ and it follows from lemma 1 that: At the optimum, none of p_i, p_j, x_i, x_j, U_i, U_j is zero then a) the budget constraint is effective and b) $MRS_{i,j} = \frac{P_i}{P_j}$.

4.2 A Problem in Welfare Economics.

In this section we study the market implications of Pareto optimality. We consider an economy with n commodities, m production processes and R consumers.

We shall retain the definitions of consumption and consumers of the previous section. Let x^r denote the vector of n commodities consumed by consumer r and let x denote the matrix whose rows are x^r, r = 1, ..., R. Let U^r denote consumer r's utility function. We shall assume that U^r depends only on x^r. Let m^r denote the income of consumer r. We assume that m^r is derived from shares in the profits of production processes; $m^r = \sum_{j=1}^{m} \beta^{rj}$ is the share of consumer r in the profits of process j and here $\beta^{rj} \geq 0$, $\sum_r \beta^{rj} = 1$.

Let $V = (V^1, \ldots, V^m)$ denote the levels at which production processes are operated. Let $g^{ij}(V^j)$ denote the net outcome, in terms of good i, when process j is operated at level V^j. We follow the convention of having $g^{ij} \leq 0$ when good i is a net input to process j at level V^j, and of having $g^{ij} > 0$ when good i is a net output of process j at level V^j. The profits of process k, if the price vector $p = (p^1, \ldots, p^n)$ prevails, is defined to be $\pi^j = \sum_i p^i g^{ij}$.

<u>Definition</u>: <u>Feasible allocation</u>: An allocation (x, v) is said to be feasible iff:

(i) $x^{ri} \geq 0$, $v^j \geq 0$,

(ii) $h^i(x, v) = \sum_r x^{ir} \geq 0$.

The definition of feasibility expresses the requirement that no more, of any commodity, is used than what is available and that consumption and process operation levels are non-negative.

<u>Definition</u>: <u>Pareto Optimum</u>: A feasible allocation $(\hat{x}, \hat{v})$ is said to be Pareto Optimal if there does not exist any other allocations (x', v') such that $^r(x'^r) \geq {}^r(\hat{x}^r)$ for all r with strict inequality for at least one r.

<u>Definition</u>: <u>Local Competitive Equilibrium</u>: A non-negative price vector p together with an allocation $(\hat{x}, \hat{v})$ is said to be a competitive equilibrium if:

Each consumer locally maximizes his utility subject to his budget constraint, i.e. there exists a neighborhood $N^r(\hat{x}^r)$ such that:

(i) $U^r(\hat{x}^r) \geq {}^r(x^r)$ for all x^r with $px^r \leq m^r$, $x^r_i \geq 0$ and $x^r \in N^r(\hat{x}^r)$.

(ii) The profits from each process are maximized, i.e.

$$pg^{\cdot,k}(\hat{v}) \geq pg^{\cdot,k}(v) \text{ for } v \geq 0$$

(iii) Demand does not exceed supply, if supply exceeds demand for a given commodity then its price is zero, i.e.

(iii.a) $h(\hat{x}, \hat{v}) \geq 0$ and

(iii.b) $ph(\hat{x}, \hat{v}) = 0$.

We now show that, under certain conditions, a pareto optimum is attainable by way of a competitive equilibrium.

<u>Proposition</u>: If 1) U and g are of class C^2. 2) $(\hat{x}, \hat{v})$ is a pareto optimum. 3) For each consumer r there exists a commodity i such that $\hat{U}^r_{x_i} > 0$.

4) $\eta^r \hat{U}^r_{x^r x^r} \eta'^r < 0$ for $\eta^r = (\eta'^r, \ldots, \eta^{nr}) \neq U$. 5) g are concave. Then there exists a price vector p and a system of shares β^{rj} such that $(\hat{x}, \hat{v})$ is a competitive

equilibrium at p.

Proof: The proof consists of essentially verifying that theorem 4 of this chapter applies.

Conditions 1) and 3) of theorem 4 are satisfied, in view of conditions 1) and 2) of our proposition. Condition 4) of the theorem is easy to verify. To verify condition 2) of the theorem it suffices to note, since U^r depends only on x^r, that condition 3 of the proposition together with the independence of g^{ik} guarantees that the matrix in definition 2) of regularity has maximum rank. Thus there exist $\alpha^r > 0$ and $\hat{\mu} \geq 0$ such that:

(1) $\hat{L}_{x^{ir}} \leq 0$, $\hat{x}^{ir}\hat{L}_{x^{ir}} = 0$, where $L = \sum_r \alpha^r U^r + \sum \hat{\mu}^i \hat{h}^i(\hat{x}, \hat{v})$.

(2) $\hat{L}_{v^j} \leq 0$, $\hat{v}^j L_{\hat{v}^j} = 0$

(3) $\hat{\mu}^i h^i(\hat{x}) = 0$.

By conditions 4) and 5) of the proposition, the function L is locally concave i.e., there exists a neighborhood of (x, v), S(x, v), such that L is concave there. By (1) and (2), using the first order sufficiency theorem, L is locally maximized at (x, v) i.e.

(4) $L(\hat{x}, \hat{v}) \geq L(x, v)$ for all $x \geq 0$, $v \geq 0$, $(x, v) \in S$.

Now take $p^i = \dfrac{1}{\sum_i \hat{\mu}^i} \hat{\mu}^i$ and take $\beta^{rj} = \beta^r = \dfrac{\sum_i p^i \hat{x}^{ir}}{\sum \hat{\pi}^j}$, where $\hat{\pi}^j = \sum_i p^i g^{ij}(\hat{v}^j)$.

Clearly, $p^i \geq 0$, $\beta^r \geq 0$ and $\sum \beta^r = \dfrac{\sum_r \sum_i p^i \hat{x}^{ir}}{\sum \hat{\pi}^j} = \dfrac{\sum \pi^j}{\sum \pi^j}$ (by (3)).

Furthermore, $m^r = \sum_j \beta^{rj}\hat{\pi}^j = \beta^r \sum \hat{\pi}^j = \sum p^i \hat{x}^{ir}$.

Relation (4) holds, in particular, if we set, on the right hand side, all x's and v's at the optimum exempt for x^{r_o} where r^o is arbitrary. Thus we have:

$$\alpha^{r_o} U^{r_o}(\hat{x}^{r_o}) - \sum_i \hat{\mu}^i \hat{x}^{ir_o} \geq \alpha^{r_o} U^{r_o}(x^{r_o}) - \sum_i \hat{\mu}^i x^{ir_o}, \text{ i.e.}$$

(5) $\alpha^{r_o}(U^{r_o}(\hat{x}^{r_o}) - U^{r_o}(x) \geq (\sum_i \hat{\mu}_i)(\sum_i p^i \hat{x}^{ir_o} - \sum p^i x^{ir_o}) = (\sum_i \hat{\mu}_i)(m^r - \sum p^i x^{ir_o})$.

By (5) $\hat{x}^{r_o}$ locally maximizes $U^{r_o}(x^{r_o})$ subject to $px^{r_o} \leq m^{r_o}$.

Since r_o is arbitrary, we have shown (i) in the definition of competitive equilibrium.

Noting that (4) holds, in particular, when we set, on the right hand side, all the x's at the optimum and all the v's at the optimum except for an arbitrary j_o. Then we get:

$$(\sum_i \hat{\mu}_i)\sum_i p^i g^{ij_o}(\hat{v}^{j_o}) \geq (\sum_i \hat{\mu}_i)(\sum_i p^i g^{ij_o}(v^{j_o})), \text{ i.e., } \hat{\pi}^{j_o} \geq \pi^{j_o},\ v^{j_o} \geq 0.$$

Note that the concavity of g guarantees that the maximum is global. Thus we have verified condition (ii) in the definition of the competitive equilibrium. Condition (iii.a) follows from feasibility, and condition (iii.b) follows from (3) above. This completes our proof.

III

VARIATIONAL PROBLEMS

CHAPTER 5

THE PROBLEM OF BOLZA WITH EQUALITY CONSTRAINTS

We shall state, here, some theorems that characterize solutions to the problem of Bolza in the calculus of variations. The proofs of some of these theorems will only be briefly outlined. The reader may refer to Bliss [8] and [9] and to Pars [40] for a more detailed presentation. By way of introduction we discuss an unconstrained problem in the calculus variation, in section 1. In section 2, we state the problem of Bolza. In section 3 we discuss first order necessary conditions and in section 4 we state the necessary conditions of Weierstrass, Clebsch, and Mayer. In section 5, we state second order sufficient conditions. In section 6, we characterize solutions to problem A'.

1. Unconstrained problem.

In this section we study characterization of constrained extrema of an integral. The results indicate the type of results expected for the constrained case and are utilized in the derivation of these results.

Let $T = [a, b]$ be a subset of the real line. Consider the set of functions $y : T \rightarrow E^n$ and the set of functions $z : T \rightarrow E^n$. Let the function $f(t, y, z) : E^{2n+1} \rightarrow E^1$ be defined for all values taken by $t \in T$, $y \in E^n$ and $z \in E^n$. Let A and B be two points in E^n. Define the set of admissible arcs M as follows:

$M = \{ y \mid$ 1) y is sectionally smooth, 2) the integral $\int_a^b f(t, y, \dot{y})\, dt$ has a finite value, 3) $y(a) = A$ and $y(b) = B\}$, where $\dot{y} = dy/dt \}$

The problem is to characterize the extreme of:

$$J[y] = \int_a^b f(t, y, \dot{y})\, dt.$$

From now on, we shall confine our attention to maxima of J[y]. We distinguish two main types of maxima: global and local.

<u>Definition 1</u>: An arc $\bar{y}$ M is said to be a global maximum of J[y] if $J[\bar{y}] \geq J[y]$, $(y \in M)$.

<u>Definition 2</u>: An arc $\bar{y}$ is said to be a local maximum of J[y] if $J[\bar{y}] \geq J[y]$, $(y \in M')$, where $M' \subset M$, $y \in M'$.

We distinguish two types of local maxima: strong and weak. The distinction is based on the definition of M'.

Definition 3: We define a strong ε-neighborhood of $\overline{y}$, $N_o(\overline{y}, \varepsilon)$ as follows: $N_o(\overline{y}, \varepsilon) = \{y \varepsilon M \mid d(y(t), \overline{y}(t)) \leq \varepsilon, (t \varepsilon T)\}$ where $d(y(\overline{t}), \overline{y}(\overline{t}))$ is the Euclidian distance, i.e. equals $\sqrt{\sum_i (y_i(\overline{t}) - \overline{y}_i(\overline{t}))^2}$, $\varepsilon > 0$.

Definition 4: A weak ε-neighborhood of $\overline{y}$, $N_1(\overline{y}, \varepsilon)$, is defined as follows:

$$N_1(\overline{y}, \varepsilon) = \{y \varepsilon N_o(\overline{y}, \varepsilon) \mid d(y(t), \overline{y}(t)) \leq \varepsilon, (t \varepsilon T)\}.$$

We are now ready to define the two types of local maxima.

Definition 5: An arc $\overline{y}$ is said to be a weak local maximum of J[y] if $J[\overline{y}] \geq J[y]$, $(y \varepsilon N_1, (\overline{y}, \varepsilon))$ for some $\varepsilon > 0$.

Definition 6: An arc $\overline{y}$ is said to be a strong local maximum of J[y] if $J[\overline{y}] \geq J[y]$, $(y \varepsilon N_o(\overline{y}, \varepsilon))$ for some $\varepsilon > 0$.

Remark 1) If $\overline{y}$ is a global maximum then it is a strong local maximum, and if $\overline{y}$ is a strong local maximum, then it is a weak local maximum. This follows from the fact that $N_1(\overline{y}, \varepsilon) \subset N_o(\overline{y}, \varepsilon) \subset M$.

1.1 First order necessary conditions.

In this section we derive a necessary condition for a weak local maximum in terms of first derivatives of J. The condition, by remark 1 above, is also necessary for strong local maxima and global maxima.

Theorem 1. If f is continuously differentiable and if $\overline{y}$ is a weak local maximum of J[y], then there exist constants C_i such that:

(0) $\overline{f}_{y_i} - \int_a^t \overline{f}_{\dot{y}_i} = C_i$, $i = 1, \ldots, n$, where $\overline{f}_{y_i} = \left.\frac{\partial f}{\partial y_i}\right| y = \overline{y}, \dot{y} = \dot{\overline{y}}$ and $\overline{f}_{\dot{y}_i} = \left.\frac{\partial f}{\partial \dot{y}_i}\right| y = \overline{y}, \dot{y} = \dot{\overline{y}}$.

Proof: Let $\eta(t): T \to E^n$ be a vector valued function, whose components are $\eta_i(t)$, such that $\eta(a) = \eta(b) = 0$ and such that $\eta_i(t)$ are sectionally smooth. For $y \varepsilon N_1(\overline{y}, \varepsilon)$ we can find α such that $y = \overline{y} + \alpha\eta$. Thus for $y \varepsilon N_1$, we may write $J[y] = \phi(\alpha)$. Note that $\phi(\alpha)$ is differentiable and that $\phi(0) = J[\overline{y}]$ is a minimum value of ϕ. Thus $\phi'(0) = 0$. The rest of the proof consists of computing $\phi'(0)$. Differentiating under the integral sign we get:

$$\phi'(\alpha) = \int_a^b \left(\sum_i f_{y_i} \eta_i + \sum_i f_{\dot{y}_i} \dot{\eta}_i \right) dt,$$

Thus

(1) $\phi'(0) = \int_a^b \left(\sum_i \bar{f}_{y_i} \eta_i + \sum_i f_{\dot{y}_i} \dot{\eta}_i \right) dt = 0.$

Integrating $\int_a^b f_{y_i} \eta_i dt$ by parts we get:

(2) $\int_a^b \bar{f}_{y_i} \eta_i dt = \eta_i \int_a^b \bar{f}_{y_i} dt \Big|_a^b - \int_a^b (\dot{\eta}_i \int_a^t \bar{f}_{y_i} dt) dt.$

But $\eta_i(a) = \eta_i(b) = 0$, and the first term of (2) vanishes. Thus we have:

(3) $\int_a^b \bar{f}_{y_i} \bar{\eta}_i dt = - \int_a^b (\dot{\eta}_i \int_a^t f_{y_i} dt) dt.$

Substituting, from (3), into (1) we have:

(4) $\int_a^b \sum_i (\bar{f}_{\dot{y}_i} - \int_a^t f_{y_i} dt) \dot{\eta}_i dt = 0.$

Define the functions $M_i(t)$ and the constants C_i as follows:

(5.1) $M_i(t) = \bar{f}_{\dot{y}_i} - \int_a^t \bar{f}_{y_i} dt,$

(5.2) $C_i = \frac{1}{b - a} \int_a^b M_i(t) dt.$

We shall show that C_i are the constraints whose existence is asserted by the theorem, i.e. they are such that $M_i = C_i$.

Since $\eta(t)$ is arbitrary, we may pick

(6) $\eta_i(t) = \int_a^t (M_i - C_i) dt$

Note that such a choice of η_i is appropriate, since η_i are sectionally smooth and since:

(7.1) $\eta_i(a) = \int_a^a (M_i - C_i) dt = 0$, and

(7.2) $\eta_i(b) = \int_a^b (M_i - C_i) dt = \int_a^b M_i dt - \int_a^b C_i dt =$

$$= \int_a^b M_i dt - (b - a) C_i$$

$$= \int_a^b M_i dt - (b - a) \frac{1}{b - a} \int_a^b M_i dt = 0$$

Now, $\dot{\eta}_i(t) = \frac{d}{dt} \int_a^t (M_i - C_i) dt = M_i - C_i.$

Thus (4) becomes

(8) $\int_a^b \sum_i M_i(M_i - C_i)dt = 0$

But (8) is equivalent to

(9) $\int_a^b \sum_i (M_i - C_i)^2 dt = 0$

The equivalence of (8) and (9) is due to:

$$\int_a^b \sum_i (M_i - C_i)^2 dt = \int_a^b M_i(M_i - C_i)dt - C_i \int_a^b (M_i - C_i)dt =$$

$$= \int_a^b M_i(M_i - C_i) - C_i \eta(b) = \int_a^b M_i(M_i - C_i)dt,$$

(by 7.2).

But the integrand and interval of integration in (9) are non-negative, thus:

(10) $\sum_i (M_i - C_i)^2 = 0.$

From (10) it follows that $M_i = C_i$. Q.E.D.

Remark: It follows from the theorem of this section that if $\bar{y}$ is a maximizing arc then

(11) $\bar{f}_{y_i} - \frac{d}{dt} \bar{f}_{\dot{y}_i} = 0.$

The equations (10) and (11) are the Euler equations.

1.2. First order sufficient conditions.

In this section we show that $\bar{y}$ is a stationary arc for $J[y]$, i.e. if it satisfies the Euler equations and if f is concave in y and $\dot{y}$ then it furnishes $J[y]$ with a weak relative maximum. Let us recall from Chapter 4 of these notes that if a function $g(x)$ is differentiable and concave, then

(1) $g(x) - g(\bar{x}) \leq \sum_i \bar{g}_i(x_i - \bar{x}_i)$, where $\bar{g}_i = \left.\frac{\partial g}{\partial x_i}\right|_{x = \bar{x}}$.

Theorem 2. If f is continuously differentiable and concave in y and $\dot{y}$, and if $\bar{y}$ is a stationary arc for $J[y]$ then $\bar{y}$ is a global maximum of $J[y]$.

Proof: Let y be an admissible arc ($y \in M$), then:

(2) $J[y] - J[\bar{y}] = \int_a^b [f(t, y, \dot{y}) - f(t, \bar{y}, \dot{\bar{y}})]dt.$

By concavity of f (and differentiability) we have:

(3) $f(t, y, \dot{y}) - f(t, \bar{y}, \dot{\bar{y}}) \leq \sum_i \bar{f}_{y_i}(y_i - \bar{y}_i) + \sum_i f_{\dot{y}_i}(\dot{y}_i - \dot{\bar{y}}_i)$, $(t \varepsilon [a, b])$.

Integrating both sides of (3) we get (in view of (2)):

(4) $J[y] - J[\bar{y}] \leq \int_a^b \left[\sum_i \bar{f}_{y_i}(y_i - \bar{y}_i) + \sum_i f_{\dot{y}_i}(\dot{y}_i - \dot{\bar{y}}_1)\right] dt.$

Since $\bar{y}$ is stationary, the right hand side may be written as:

(5) $\int_a^b \left[\sum_i \frac{d}{dt} \bar{f}_{\dot{y}_i}(y_i - \bar{y}_i) + \sum_i f_{\dot{y}_i}(\dot{y}_i - \dot{\bar{y}}_i)\right] dt.$

Integrating the first term by parts, the expression (5) becomes:

(6) $\sum_i (y_i - \bar{y}_i) f_{\dot{y}_i} \Big|_a^b - \int_a^b \sum_i f_{\dot{y}_i}(\dot{y}_i - \dot{\bar{y}}_i) dt + \int_a^b \sum_i f_{\dot{y}_i}(\dot{y}_i - \dot{\bar{y}}) dt/$

Thus we have (in view of (4), (5) and (6)):

(7) $J[y] - J[\bar{y}] \leq \sum_i (y_i - \bar{y}) \bar{f}_{\dot{y}_i} \Big|_a^b.$

But y is in M, thus $y(a) - \bar{y}(a)$ and $y(b) = \bar{y}(b)$, i.e. the right hand side of (7) is zero. Thus:

$$J[y] - J[\bar{y}] \leq 0,$$

and since y is an arbitrary arc in M, $\bar{y}$ is a global maximum arc for J[y]. Q.E.D.

1.3 The necessary condition of Weierstrass.

We define the Weierstrass excess function at maximal arc $\bar{y}$ as the change in the value of the integrand f when $\dot{\bar{y}}$ is replaced by an arbitrary vector p of real numbers:

$$E(t, \bar{y}, \dot{\bar{y}}, p) = f(t, \bar{y}, \dot{\bar{y}}) - f(t, \bar{y}, p) - (p - \dot{\bar{y}})\bar{f}_{\dot{y}}.$$

Theorem 3. If f is of class C^2 and if $\bar{y}$ provides a weak local maximum for J[y] in a weak neighborhood $N_1(\bar{y})$, then: $E(t, \bar{y}, \dot{\bar{y}}, p) \geq 0$ for all p and for all $t \varepsilon [a, b]$.

Proof: The proof is by contradiction. We shall assume that E is negative for some p_o at $t = t_o$, $t_o \varepsilon [a, b]$. Then we may find an arc $\tilde{y}(t) \varepsilon N_1$ such that $J[\tilde{y}] > J[\bar{y}]$.

Define $\tilde{y}(t; h) = \begin{cases} \bar{y}(t), & t \varepsilon [t_o, b] \\ (t - t_o).p_o + \bar{y}(t_o), & t \varepsilon [t_o - h, t_o] \\ (t - a). k + \bar{y}(t) & t \varepsilon [a, t_o - h], \end{cases}$

where h denotes a real number with $t_o - h > a$ and where k is a n-vector function defined by:

(1) $\overline{y}(t_o - h) + (t_o - a - h).k(h) = \overline{y}(t_o) - p_o h.$

Using the definition of $\tilde{y}(t; h)$ we shall write the difference between $J[\overline{y}]$ and $J[\tilde{y}]$ as a function $\omega(h)$, of a real number h. We then show that $\omega(0) = 0$ and $\omega'(0) > 0$ thus showing that ω is a monotone increasing in a neighborhood of $h = 0$ and that for $h > 0$, in that neighborhood, $\omega(h) > 0$. But then $J[\tilde{y}] - J[\overline{y}] = \omega(h) > 0$, which would complete our proof.

We write $\omega(h) = J[\tilde{y}] - J[\overline{y}]$. Clearly $\omega(0) = 0$, since $\tilde{y}(t, 0) = \overline{y}(t)$. But,

$$\omega(h) = \int_a^b f(t, \tilde{y}(t, h), \dot{\tilde{y}}(t, h)) - f(t, \overline{y}, \dot{\overline{y}})dt = \int_a^{t_o - h} f(t, \tilde{y}, \dot{\tilde{y}}) - f(t, \overline{y}, \dot{\overline{y}})\, dt +$$

$$+ \int_{t_o - h}^{t_o} f(t, \tilde{y}, \dot{\tilde{y}}) - f(t, \overline{y}, \dot{\overline{y}})dt + \int_{t_o}^b f(t, \tilde{y}, \dot{\tilde{y}}) - f(t, \overline{y}, \dot{\overline{y}})dt.$$

The last term is zero, by (1) and $\dot{\tilde{y}} = \begin{cases} p_o & t \in (t_o - h, t_o) \\ \dot{\overline{y}} + k, & t \in (a, t_o - h) \end{cases}$

and we write:

(2) $$\omega(h) = \int_a^{t_o - h} f(t, \overline{y} + (t - a). k, \dot{\overline{y}} + k)dt + \int_{t_o - h}^{t_o} f(t, \overline{y}(t_o) + (t - t_o). p_o, p_o)dt - \int_a^{t_o} f(t, \overline{y}, \dot{\overline{y}})dt.$$

Differentiating under the integral sign in (2) with respect to h we get:

(3) $$\omega'(h) = \int_a^{t_o - h} [(t - a).f_y(t, \tilde{y}, \dot{\tilde{y}})k_h + f_{\dot{y}}k_h]dt -$$

$$- f(t_o - h, \overline{y}(t_o - h) + (t_o - h - a).k, \dot{\overline{y}}(t_o - h) + k) +$$

$$+ f(t_o - h, \overline{y}(t_o) - h.p_o, p_o) \text{ where } k_h = \frac{d}{dh} k.$$

Evaluating (3) at $h = 0$, we have:

(4) $$\omega'(0) = \int_0^{t_o} (t - a). (\overline{f}_y + \overline{f}_{\dot{y}})k_h^o + f(t_o, \overline{y}(t_o), p_o) - f(t_o, \overline{y}(t_o), \dot{\overline{y}}(t_o))$$

where $k_h^o = k_h \big|_{h = 0}$.

By the maximality of $\overline{y}$ we get, by theorem 1:

$$f_y = \frac{d}{dt} f_{\dot{y}}.$$

Thus we may write: $\int_a^{t_o} [(t - a)\, \overline{f}_y + \overline{f}_{\dot{y}}]dt$ as $\int_a^{t_o} [(t - a). \frac{d}{dt} f_{\dot{y}}]\, dt + \int_a^{t_o} \overline{f}_{\dot{y}}dt.$

Integrating the first term by parts we get:

$$\int_a^{t_o} (t - a) \frac{d}{dt} f_{\dot{y}} dt = (t - a).\bar{f}_{\dot{y}} \Big|_a^{t_o} = (t_o - a).\bar{f}_{\dot{y}}.$$

So we have:

$$(5) \quad \int_a^{t_o} [(t - a).\bar{f}_y + \bar{f}_{\dot{y}}]\, dt = (t - a).\bar{f}_{\dot{y}} \Big|_a^{to} = (t_o - a).\bar{f}_{\dot{y}}.$$

Differentiating both sides of (1) with respect to h we have:

$$- \dot{\bar{y}} (t_o - h) - k + (t_o - a - h)k_h = -p_o.$$

Evaluating at $h = 0$, we get:

$$(t_o - a)k_h^o = \dot{\bar{y}}(t_o) - p_o + k(0).$$

But, evaluating (1) at $h = 0$, we have:

$$y(t_o) + (t_o - a)\, k(0) = y(t_o),$$

and $k(0) = 0$, since $a - t_o > 0$. Thus

$$(6) \quad (t_o - a).\ k_h^o = \dot{\bar{y}}(t_o) - p_o.$$

Substituting from (5) and (6) into (4) we have:

$$\begin{aligned} \omega'(0) &= (t_o - a)\, k_h^o\, \bar{f}_{\dot{y}} + f(t_o, \bar{y}(t_o), p_o) - f(t_o, \bar{y}(t_o), \dot{\bar{y}}(t_o)) \\ &= (\dot{\bar{y}}(t_o) - p_o)\, \bar{f}_{\dot{y}} + f(t_o, \bar{y}(t_o), p_o) - f(t_o, \bar{y}(t_o), \dot{\bar{y}}(t_o)) \\ &= - E(t_o, \bar{y}(t_o), \dot{\bar{y}}(t_o), p_o) > 0. \end{aligned}$$

Since ω' is continuous, there exists a neighborhood $S(0)$ of $h = 0$ where $\omega'(h) > 0$, choose $h \in S(0)$ such that $\tilde{y} \in N_1$. Q.E.D.

1.4 Legender necessary conditions.

Theorem 4. If f and $\bar{y}$ are as in theorem 3, then $\eta'\bar{f}_{\dot{y}\dot{y}}\eta \leq 0$ for all η for $t \in [a, b]$.

Proof: The proof is by contradiction and it uses the construction of theorem 3. Suppose there exists $t_o \in [a, b]$ and $0 \neq \eta_o \in E^n$ such that $\eta_o' f_{\dot{y}\dot{y}} (t_o, \bar{y}(t_o), \dot{\bar{y}}(t_o))\eta_o$ is positive. Then using Taylor's theorem of the second order we may show that the Weierstrsss E function is positive at t_o for some value of p. We then may construct an arc in N_1 which gives J a larger value than $J[\bar{y}]$, as we did in the proof of theorem 3, which would contradict our hypothesis.

1.5 Sufficient conditions.

In this section we show, under some additional assumptions, that the Weierstrass condition is also sufficient. We also prove a second order sufficient condition. In preparation for this we start by some notations, definitions and preliminary results.

Definition: We define a stationary arc as an arc that satisfies the Euler equations of theorem 1.

Notation: Let $S = \{\text{the set of stationary arcs}\}$.

Definition: (Fields of a functional): Let $G \subset E^{n+1}$ be the set of values taken by $(t, y(t))$. Consider D G and the function $q(t, y) : D \to E^n$. The pair $\langle D, q\rangle$ is said to be a field of $J[y]$ iff:

1) $q \in C^1$ on D,

2) The Hilbert integral, $I_c = \int_c [(f(t, y, q) - q.f_{\dot y})]\, dt + f_{\dot y}(t, y, q).\, dy$, where $dy = (dy_1, dy_2, \ldots, dy)$ and where the $a \cdot b$ denotes $\sum_{i=1}^{n} a_i b_i$,

is independent of the path of integration.

Definition: (A set covered by a field, slope): If $\langle D, q\rangle$ is a field, then we say that D is covered by a field and that q is the slope function of the field.

Definition: Let $J = \{y \mid y = \phi(t, \beta),\ \beta \in E^n,\ \dot y = q(t, y)\}$. J is called the trajectory of the field.

Remark: By existence of theorems of solutions to differential equations, y is unique for any given β and ϕ, together with ϕ_t, is in C^1 with respect to β for some region R in E^{n+1}.

Lemma 1: All arcs in a field are stationary.

Proof: Recall that a necessary condition for a line integral to independent of path is that the cross partials be equal. Since the Hilbert integral is independent of path, by definition of a field we have:

$$(1)\quad \frac{\partial}{\partial y_i} f_{\dot y_k}(t, y, q) = \frac{\partial}{\partial y_k} f_{\dot y_i}(t, y, q), \text{ and}$$

$$(2)\quad \frac{\partial}{\partial y_k}(f(t, y, q) - q.f_{\dot y}) = \frac{\partial}{\partial t} f_{\dot y_k}(t, y, q).$$

Performing the differentiation in (2) we have:

$$(3)\quad f_{y_n} + f_{\dot y}.q_{y_k} - q_{y_k} f_{\dot y} - \sum_i \frac{\partial}{\partial y_k} f_{\dot y_i} = f_{\dot y_k t} + \sum_i f_{\dot y \dot y_i} \frac{\partial q_i}{\partial t}$$ where q_{y_k} is the n-vector of partial derivatives of q with respect to y_k.

Thus, by (1),

$$(4)\quad f_{y_k} = f_{\dot{y}_k t} + \sum_i q_i \frac{\partial}{\partial y_i} f_{\dot{y}_k} + f_{\dot{y}\dot{y}}$$

But $\frac{\partial}{\partial y_i} f_{\dot{y}_k} = f_{\dot{y}_k y_i} + \sum_j f_{\dot{y}_k \dot{y}_j} \frac{\partial q_j}{\partial y_i}$, and

$$(5)\quad f_{y_k} = f_{\dot{y}_k t} + \sum_i f_{\dot{y}_k \dot{y}_i} q_i + \sum_i f_{\dot{y}_k \dot{y}_i} \left(\frac{\partial q_i}{\partial t} + \sum_j j \frac{\partial q_i}{\partial y_j}\right).$$

But, (5) holds for $\dot{y}_i = q_i$ and for

$$(6)\quad \ddot{y}_i = \frac{\partial q_i}{\partial t} + \sum_j \frac{\partial q_i}{\partial y_j} \frac{dy_j}{dt} = \frac{\partial q_i}{\partial t} + \sum_j \frac{\partial q_i}{\partial y_j} q_j.$$

By (5) and (6) we have:

$$f_{y_k} = f_{\dot{y}_k t} + \sum_i f_{\dot{y}_k y_i} \dot{y}_i + \sum_i f_{\dot{y}_k \dot{y}_i} \ddot{y}_i = \frac{d}{dt} f_{\dot{y}_k}.$$

Q.E.D.

Lemma 2. <u>Let S(B) be an n-parameter family of stationary arcs for J such that</u>:

1) $y = \phi(t, \beta)$, $\phi : R \to D$,

2) $|\phi_\beta| \neq 0$,

3) R is simply connected.

<u>Then <S(B), $\dot{y}$> is a field iff</u>:

$$4)\quad \sum_{i=1}^{n} \frac{\partial y_i}{\partial \beta_s} \frac{\partial \nu_i}{\partial \beta_r} - \frac{\partial y_i}{\partial \beta_r} \frac{\partial \nu_i}{\partial \beta_s} \overset{t}{\equiv} 0,$$

for r, s going from 1 to n and where $\nu_i = f_{\dot{y}_i}(t, y, \dot{y})$.

Proof: The implicit function theorem applies and we may solve 1) for β and get:

$$\beta_i = \Psi_i(t, y)$$

We shall show that the slope of the field is $\dot{y} = \phi_t(t, \beta) = \phi_t(t, \Psi(t, y) = q$. We shall show that this choice of q is proper iff 4) holds by showing that 4) is necessary and sufficient for the Hilbert integral to be independent of the path of integration. Let L be an arc in S.

$$I_L = \int_L [(f(t, \phi, \phi_t) - q.f_{\dot y}(t, \phi, \phi_t))\, dt + f_{\dot y} d\phi]$$

$$= \int_L (f + \phi_t . f_{\dot y})\, dt - \int_L f_{\dot y}\phi_t dt + \int_L f_{\dot y} d\phi$$

$$= \int_L f dt + \int_L f_{\dot y} . \phi_t dt - \int_L f_{\dot y} . \phi_t dt + \int_L f_{\dot y} . d\phi$$

$$= \int_L f dt + \sum_k \sum_i f_{\dot y_i} \omega^i_{\beta_k} d\beta_k .$$

Since R is simply connected, a necessary and sufficient condition for the last integral to be independent of path is that the cross partials of the integral be equal i.e., that:

(1) $\frac{\partial}{\partial \beta_k} f = \frac{d}{dt} f_{\dot y_i} \phi^i_{\beta_k}$ and

(2) $\frac{\partial}{\partial \beta_r} (\sum_i f_{\dot y_i} \phi^i_{\beta_s}) = \frac{\partial}{\partial \beta_s} (\sum_i f_{\dot y_i} \phi^i_{\beta_r})$.

Performing the differentiation in (1) we have:

(1)' $\sum_i f_{y_i} \phi^i_{\beta_k} + \sum_i f_{\dot y_i} \phi^i_{t\beta_k} = \sum_i f_{\dot y_i} \omega^i_{\beta_{kt}} + \sum_i \omega^i_{\beta_k} \frac{d}{dt} f_{\dot y_i}$,

which is equiv lent to:

(1)" $\sum_i (f_{y_i} - \frac{d}{dt} f_{\dot y_i}) \phi^i_{\beta_t} = 0$,

which is true by the stationarity of L. Thus independence of path is equivalent to (2), which may be written as:

(2)' $\sum_i \phi^i_{\beta_s} \frac{\partial}{\partial \beta_r} f_{\dot y_i} = \sum_i \phi^i_{\beta_r} \frac{\partial}{\partial \beta_s} f_{\dot y_i}$.

Noting that $\phi^i_{\beta_j} = \frac{\partial y_i}{\partial \beta_i}$, we conclude that (2)' is the relation 4) in the lemma and the proof is complete.

Note: The terms in 4) are known as Lagrange brackets.

Theorem 5. (Sufficient condition of Weierstrass) If 1) $f \in C^2$. 2) $\bar y$ is stationary. 3) $\bar y$ can be imbedded in a field, i.e. there exists a set D containing $\bar y$ that is covered by a field with $\bar y$ as one of its trajectories. 4) $E(t, y, q(t, y), p) \leq 0$ on D, with p finite and $y \in D$. Then $\bar y$ provides $J[y]$ with a local maximum on D.

Proof: Consider $\tilde y(t) \in D \cap M$. To show: $J[\tilde y] - J[\bar y] \leq 0$. Along $\bar y$ we have, the Hilbert integral, $I_{\bar{y}} = \int_{\bar{y}} (f, (t, \bar{y}, \dot{\bar{y}}) - \dot{\bar{y}}.f_{\dot{y}}) dt + \dot{y}f_{\dot{y}} dt = J[\bar{y}]$. Since $\tilde{y}$ is in the field, $I_{\tilde{y}}$ is independent of path, and

$$J[\bar{y}] = I_{\bar{y}} = I_{\tilde{y}} = \int_a^b [f(t, \tilde{y}, \dot{\tilde{y}}) - (q(t, \tilde{y}) - \dot{\tilde{y}}).f_{\dot{y}} (t, \tilde{y}, q(t, \tilde{y}))]dt.$$

Thus

$$J[\tilde{y}] - J[\bar{y}] = \int_a^b [f(t, \tilde{y}, \dot{\tilde{y}}) - f(t, \tilde{y}, q(t, \tilde{y}) - (\dot{\tilde{y}} - q(t, \tilde{y})f_{\dot{y}} (t, \tilde{y}, q(t, \tilde{y}))]dt,$$

which is non-positive by assumption 4) of the theorem. Q.E.D.

Theorem 6. (Legender sufficient condition.) <u>If f and $\bar{y}$ are as in theorem 5 and if $\eta\bar{f}_{\dot{y}\dot{y}}\eta^* < 0$ for $\eta \neq 0$, then $\bar{y}$ provides a weak local maximum for $J[y]$.</u>

Proof: The proof consists of showing that there exists a weak neighborhood of $\bar{y}$ such that the values of the arcs in that neighborhood are in D and such that the Weierstrass inequality holds there. The theorem, then, follows from theorem 5.

2. Statement of the Problem of Bolza.

Let T be a subset of the non-negative half line. Let M be the set of sectionally smooth functions from T into E^n.

Problem A: Among the arcs in M, maximize

(1) $J^o[y] = h^o(t_o, t_1, y_o, y_1) + \int_{t_o}^{t_1} L^o(t, y, \dot{y})dt$, where $t_o < t_1$ are in T,

$y_o = y(t_o)$, $y_1 = y(t_1)$, subject to:

(2) $\omega^\alpha(t, y, \dot{y}) = 0$, $\alpha = 1, \ldots, m$, $m < n$,

(3) $h^\beta(t_o, t_1, y_o, y_1) = 0$, $\beta = 1, \ldots, p$, $p < 2n + 1$.

We shall say that $\bar{y}$ provides a global, weak local or strong local solution to problem A if $\bar{y}$ provides a global, weak local or strong local maximum of (1) subject to (2) and (3).

The problem we stated is equivalent* to the following problem:

Problem A': Maximize J[y], where J[y] is defined in (1), subject to (2) and to

(3)' $J^\beta[y] = h^\beta(t_0, t_1, y_0, y_1) + \int_{t_0}^{t_1} L^\beta(t, y, \dot{y}) dt$.

Problem A is a special case of problem A', as could be seen by setting $L^\beta = 0$ in (3)'.

(*) In the sense that each problem may be formulated as a special case of the other.

Problem A' is a special case of problem A, as could be seen by introducing the auxiliary variable $z^\beta(t)$ as follows:

(2)' $\dot{z}^\beta - L^\beta(t, y, \dot{y}) = 0$,

augmenting the constraints (2) by (2)' and writing (3)' as:

$$h^\beta(t_0, t_1, y_0, y_1) + z^\beta(t_1) - z^\beta(t_0) = 0$$

which is in the form of (3).

We shall outline the proofs of some of our theorems for problem A and state the corresponding theorems for problem A', indicating the proof for some of these theorems.

We now introduce some definitions and notation. The variation of an arc y, due to a change in a parameter b, where b is a real vector with one or more components is denoted by Sy and the variations of the points t_0 and t_1 are denoted, respectively by St_0 and St_1. The variation along an arc is a "shift" in the arc. Let $\overline{\Gamma} = (\overline{t}_0, \overline{t}_1, \overline{y})$ be a solution to problem A. We shall denote the variation along $\overline{\Gamma}$ by $p = (\tau_0, \tau_1, \eta)$, where p is an n + 2 vector if b has one component and p is an s x n + 2 matrix if b has S components. In case p is a matrix, we shall denote the rows of p by p^σ, $\sigma = 1, \ldots, S$.

<u>Definition</u>: <u>An s-parameter family of arcs</u>. Let b be as above and let $t_0(b)$ and $t_1(b)$ be real valued functions of b. Let y(t, b) be an n-vector valued function of b for $t \in [t_0(b), t_1(b)]$. We say that y(t, b) is an S-parameter family of arcs.

<u>Definition</u>: <u>Admissible representations</u>. An S-Parameter family of arcs is said to be an admissible representation starting at $\overline{y}$ if:

1) $y(t, 0) = \overline{y}$, $t_0(0) = \overline{t}_0$, $t_1(0) = \overline{t}_1$.

2) i) $St_0 = t_{0b}(0)db = \tau_0 db$.

ii) $St_1 = t_{1b}(0)db = \tau_1 db$

iii) $Sy_i = y_{ib}(t, 0)db = \eta_i(t)db$,

where the second subscript denotes differentiation with respect to b and where db is an s-vector.

3) $\phi^\alpha(t, y(t, b), \dot{y}(t, b)) = 0$

4) $\Phi^\alpha(t, \eta, \dot{\eta}) = \omega^\alpha_y \eta^* + \omega^\alpha_{\dot{y}} \dot{\eta}^* = 0$,

where η is an s x n matrix. Equations 4) are called the equations of variation along $\bar{y}$.

3. First order necessary conditions.

In this section we state the multiplier rule for the problem of Bolza and outline the proof. The outline consists of a sequence of assertions, whose proofs may be found, e.g., in Bliss [ ] or [ ]

Theorem 7. If: 1) h. L^o and ϕ are in C^2. 2) $\bar{\Gamma}$ is a solution to problem A. 3) The matrices $[\bar{\omega}_{\dot{y}}]$ and $[h_{t_o} h_{t_1} h_{y_o} h_{y_1}]$ have maximal rank. Then there exist constants $(\lambda_o, \gamma) \neq 0$ with p + 1 components, $\lambda_o \geq 0$, and an m-vector valued function $\lambda(t)$ such that $(\lambda_o, \lambda(t))$ is never zero and:

(1) Euler-Lagrange equations: $\frac{d}{dt} \bar{F}_{\dot{y}} = \bar{F}_y$ where $F = \lambda_o L^o + \lambda\phi$.

(2) $\frac{d}{dt} (\bar{F} - \dot{\bar{y}} \bar{F}_{\dot{y}}) = \frac{\partial}{\partial t} \bar{F}$.

(3) Transversality conditions:

i) $G_{t_o} + \dot{\bar{y}}_o G_{y_o} - \bar{F}\big|_{t = t_o} = 0$

ii) $G_{t_1} + \dot{\bar{y}}_1 G_{y_1} + \bar{F}\big|_{t = t_1} = 0$

iii) $G_{y_o} - F_{\dot{y}}\big|_{t = t_o} = 0$

iv) $G_{y_1} + F_{\dot{y}}\big|_{t = t_1} = 0,$

where $G = \lambda_o h^o + \gamma h$, $\dot{\bar{y}}_r = \dot{\bar{y}}(t_r)$, $(r = 0, 1)$, $\bar{F}\big|_{t = t_r} = \bar{F}$ with all argument evaluated at t_r and $F_{\dot{y}}\big|_{t = t_r}$ is defined similarly.

Outline of the proof:

1. Condition (2) follows from condition (1).

2. Condition (1) is implied by:

(1)' There exists a constant n-vector such that $\bar{F}_{\dot{y}} = \int_{t_o}^{t} F_y + C$.

3. Condition (3) is equivalent to:

dt_o
dt_1
dy_o
dy_1

(3)' $[(\bar{F} - \dot{y}\bar{F}y)dt + dyF_{\dot{y}}]\Big|_0^1 + dG \equiv 0$

To see this, consider the transformation:

$$dt_o = \tau_o db,\ dt_1 = \tau_1 db,\ dy_{io} = \dot{y}_{io} d\tau_o + \eta_{io} db,\ dy_{i1} = \dot{y}_{i1} d\tau_1 + \eta_{i1} db.$$

4. By 1, 2, and 3 it suffices to prove (1)' and (3)'.

5. There exists a (p + 1)-parameter family which is an admissible representation, starting at $\bar{y}$. This is accomplished by adjoining, to the constraints $\phi^\alpha = 0$, n - m equations $\phi^\sigma(t, y, \dot{y}) = z^\sigma$, where $z^\sigma(t)$ are auxiliary variables. This is done so that $\begin{bmatrix} \phi^\alpha_{\dot{y}} \\ \phi^\sigma_{\dot{y}} \end{bmatrix}$ has rank m. We then solve the system for $\dot{y} = \Psi(t, y, z)$. Then consider:

$\dot{y} = \Psi(t, y, z + b^\sigma\zeta)$, $\sigma = 1, \ldots, p + 1$. Solutions depend on t_o, y_o, y_1 and on b. $y = y(t, b)$ obtained this way is an admissible representation. Substituting into $J[y]$ we have: $b = 0$ solves the following problem:

Maximize $J(b)$ subject to $h^\beta(b) = 0$, $\beta = 1, \ldots, p$. This is a problem in E^n, and theorem 1 of chapter 2 applies. Thus, there exist constants $(\lambda_o, \gamma) \neq 0$ such that:

$$(1)\quad \frac{\partial}{\partial b^\sigma} (\lambda_o J + \gamma h(b))\Big|_{b = 0} = 0.$$

6. Denoting $\frac{\partial}{\partial b} J$ evaluated at $b = 0$ by J_1, we have:

$$(2)\quad J_1 = (\bar{h}^o_{t_o} + \bar{h}^o_{y_o} \dot{y}_o)\ \tau_o + (\bar{h}^o_{t_1} + \bar{h}^o_{y_1} \dot{\bar{y}}_1)\tau_1 + \bar{h}^o_{y_o} \eta_o + \bar{h}^o_{y_1} \eta_1 + \bar{L}^o_1\ \tau_1 - \bar{L}^o_o\ \tau_o + \int_{\bar{t}_o}^{\bar{t}_1} (f_y\eta + f_{\dot{y}}\dot{\eta})\ dt.$$

7. Denoting $\frac{\partial}{\partial b} h^\beta$ evaluated at $b = 0$ by h^β_1, we have:

$$(3)\quad h^\beta_1 = (\bar{h}^\beta_{t_o} + \bar{h}^\beta_{y_o}\ \dot{y}_o)\tau_o + \bar{h}^\beta_{t_1} + \bar{h}^\beta_{y_1}\ \dot{\bar{y}})\tau_1 + \bar{h}^\beta_{y_o}\eta_o + \bar{h}^\beta_{y_1}\eta_1.$$

8. Recall that our variations satisfy:

$$(4.\text{i})\quad \Phi^\alpha = \phi^\alpha_y\eta + \omega^\alpha_{\dot{y}}\dot{\eta} = 0$$

$$(4.\text{ii})\quad \Phi^\sigma = \bar{\phi}^\sigma_y\eta + \bar{\phi}^\sigma_{\dot{y}}\dot{\eta} - \zeta^\sigma = 0,\ \text{where } \zeta^\sigma \text{ is the variation of } z^\sigma.$$

9. Multiplying equations (4) by some functions $\ell^\alpha(t)$ and $\ell^\sigma(t)$ and integrating, we get:

(5) $\int_{\bar{t}_o}^{\bar{t}_1} \ell^I\Phi^I + \ell^{II}(\Phi^{II} - \zeta^{II})\ dt = 0$, where the components of ℓ^I and Φ^I and

where the components of ℓ^{II} and Φ^{II} are ℓ^σ and Φ^σ.

10. Define $\tilde{F} = \lambda_o L^o + \ell\phi$, where $\ell = (\ell^I, \ell^{II})$ and $\phi = (\phi^I, \phi^{II})$. Take ℓ to be solutions of:

(6) $\tilde{F}_{\dot{y}} = \int_{t_o}^{t} \tilde{F}y \, dt + C$, where C is an arbitrary constant vector.

11. We may take $\lambda^\alpha(t) = \ell^\alpha(t)$.

12. $\ell^{II}(t) = 0$, and $\tilde{F} = F$. This would establish (1)'. The assertion is proved as follows: multiply (2) by λ_o and add (5) to the right hand side, and use integration by parts, (1), (6) and the arbitrariness of ζ to show that $\ell^{II} = 0$.

13. The transversality condition (3)' follows from (1).

5. Weierstrass and second order necessary conditions.

Define $E(t, y, \dot{y}, \lambda_o, \lambda, p) = F(t, y, p, \lambda_o, \lambda) - F(t, y, \dot{y}, \lambda_o, \lambda) - (p - \dot{y}) F_{\dot{y}}(t, y, \dot{y}, \lambda_o, \lambda)$.

Theorem 8. (Weierstrass necessary condition) *If all the hypotheses of theorem 7 are satisfied then $E(t, \bar{y}, \dot{\bar{y}}, p, \lambda_o, \lambda) \leq 0$ for $(\bar{y}, p)$ satisfying $\phi^\alpha(t, y, \dot{y}) = 0$.*

Theorem 9. (Clebsch second order necessary conditions) *If all the hypotheses of theorem 7 are satisfied then $\pi \bar{F}_{\dot{y}\dot{y}} \pi^* \leq 0$ for all π with $\phi^\alpha_{\dot{y}} \pi^* = 0$, $\alpha = 1, \ldots, m$.*

Remark: Theorems (8) and (9) were proved by McShane [36] using separation of convex sets.

Before we state our next theorem we define the concept of normality and state a necessary and sufficient condition for normality.

Definition: $\bar{\Gamma}$, a solution to problem A, is said to be normal if it satisfies the first order necessary conditions with $\lambda_o = 1$ and with a unique multiplier vector, (λ, γ).

Lemma. *$\bar{\Gamma}$ is normal iff it satisfies the rank condition in the sense of definition D.13 of Chapter 1.*

Theorem 10. (Jacobi-Myer-Bliss necessary conditions) *If, in addition to the assumptions of theorem 7, the rank condition is satisfied, then there exist multipliers as in theorem 7 with $\lambda_o = 1$ such that:*

$$\overline{J}_2 = d^2G(\overline{t}_o, \overline{t}_1, \overline{y}_o, \overline{y}_1; \tau_o, \tau_1, \eta_o, \eta_1) + [(\overline{F}_t - \overline{F}_y\dot{\overline{y}})\tau_r^2 +$$

$$+ 2\overline{F}_y\eta(t_r)\tau_r]_{r=0}^{r=1} + \int_{\overline{t}_o}^{\overline{t}_1} \eta\overline{F}_{yy}\eta^* + 2\eta\overline{F}_{y\dot{y}}\dot{y}^* + \eta F_{\dot{y}\dot{y}}\dot{\eta}^* \, dt \leq 0, \text{ for } (\tau_o, \tau_1, \eta) = p,$$

$$\Phi^\alpha = 0 \text{ and } \overline{H}^\beta = (h^\beta_{t_o} + \dot{y}_o h^\beta_y)\tau_o + h^\beta_{y_o}\eta_o + (h^\beta_{t_1} + \dot{y}_1 h^\beta_{y_1})\tau_1 + h^\beta_{y_i}\eta_1 = 0.$$

5. Sufficient conditions

Theorem 11. (First order sufficient conditions) If 1) L^o, h^o, h and ϕ are of class C^2. 2) $\overline{\Gamma}$ satisfies the first order necessary conditions, Euler-Lagrange Equations, with $\lambda_o > 0$. 3) The function F and G, defined in theorem 7 are concave. Then $\overline{\Gamma}$ is a global solution to problem A.

Theorem 12. (Second order sufficient conditions) If, in addition to assumptions 1) and 2) of theorem 11, we have $\overline{J}_2 < 0$ for every $p \neq 0$ with $\Phi^\alpha = 0$ and $H^\beta = 0$. Then $\overline{\Gamma}$ is a weak local solution of problem A.

6. Characterization of problem A'.

(a) Necessary conditions: If 1) h^o, h, L^o, L and ϕ are of class C^2. 2) $\overline{\Gamma}$ is a solution to problem A'. 3) The matrices $\begin{matrix}\overline{\phi}_{\dot{y}}\\ L_{\overline{y}}\end{matrix}$, $[\overline{h}_{t_o}, \overline{h}_{t_1}, \overline{h}_{y_o}, \overline{h}_{y_i}]$ have maximal ranks. Then there exists constants $(\lambda_o, \gamma^1) \neq 0$ and functions $\lambda'(t)$ with (λ_o, λ^1) never zero such that:

(1) Euler-Lagrange Equations: $\frac{d}{dt}\overline{F}_{\dot{y}} = \overline{F}'_{y'}$, where $F' = \lambda^o L^o + \gamma' L + \lambda\phi$.

(2) $\frac{d}{dt}(\overline{F}' - \dot{\overline{y}}\overline{F}'_{\dot{y}}) = \frac{\partial}{\partial t}\overline{F}'$

(3) Transversality conditions: $[(\overline{F}' - \dot{y}\overline{F}'_{\dot{y}})dt + dy\overline{F}'_{\dot{y}}]\Big|_0^1 + dG' \equiv 0$, where $\begin{matrix} dt_o \\ dt_o \\ dy_o \\ dy_1 \end{matrix}$

$G' = \lambda_o h^o + \gamma' h$.

(4) Weierstrass condition. Define $E'(t, y, \dot{y}, p, \lambda_o, \lambda') = F'(t, y, p, \lambda_o, \lambda') - F'(t, y, \dot{y}, \lambda_o, \lambda') - (p - \dot{y})F'_{\dot{y}}(t, y, \dot{y}, \lambda)$. $E(t, \overline{y}, \dot{\overline{y}}, p, \lambda_o, \lambda') \leq 0$ for y, p satisfying $\phi^\alpha(t, y, p) = 0$.

(5) Clebsch condition. $\pi\overline{F}'_{\dot{y}\dot{y}}\pi^* \leq 0$ for π satisfying $\overline{\phi}_{\dot{y}}\pi^* = 0$.

(6) Mayer-Jabobi-Bliss condition. If $\overline{\Gamma}$ is normal then $\overline{J}'_2 \leq 0$ for each p with

$\overline{\Phi}^\alpha = 0$, $\overline{H}'^\beta = \overline{H}^\beta + \int_{t_o}^{\overline{t}_1} (L^\beta_y \eta + L^\beta_{\dot{y}} \dot{\eta})\, dt = 0$, where J'_2 is analogous to J_2 with F and G replaced by F' and G'.

(b) Sufficient conditions: Suppose h^o, h, L^o, L and ϕ are of class C^2 and conditions 1) and 3) of (a) above hold at $\overline{\Gamma}$. Then we have:

(b.1) If $\lambda_o > 0$ and F' and G' are concave then $\overline{\Gamma}$ is a global solution to problem A'.

(b.2) If $\overline{J}'_2 < 0$ for every $p \neq 0$ with $\overline{\Phi}^\alpha = 0$ and $\overline{H}'^\beta = 0$, then $\overline{\Gamma}$ is a weak local solution to problem A'.

We introduce the auxiliary variables z^β as follows:

$$\dot{z}^\beta - L^\beta(t, y, \dot{y}) = 0, \; \beta = 1, \ldots, p.$$

Thus, we have a problem of type A of Maximizing J subject to $\phi^\alpha = 0$, $\dot{z}^\beta - L^\beta = 0$, $h^\beta + z^\beta_1 - z^\beta_o = 0$. Define the, path, Lagrangian $F = \lambda_o L^o + \lambda^I \phi + \lambda^{II} L$ and the end point Lagrangian $G = \lambda_o L^o + \gamma h$. By the Euler-Lagrange conditions for problem A, $\frac{d}{dt} \overline{F}_{\dot{y}} = \overline{F}_y$ and $\frac{d}{dt} \overline{F}_{\dot{z}} = \overline{F}_z = 0$. But $\overline{F}_{\dot{z}} = \lambda^{II}$, hence $\dot{\lambda}^{II} = 0$, and λ^{II} is a constant vector. By transversality conditions for problem a, $\lambda^{II} = \gamma$. Thus taking $\lambda' = \lambda^I$ and $\gamma' = \gamma$ the characterization of problem A' follows from that of problem A.

CHAPTER 6

THE PROBLEM OF BOLZA WITH EQUALITY-INEQUALITY CONSTRAINTS

In this chapter, we study the problem of Bolza with added inequality constraints. We shall use the theorems of chapter 5 to prove our characterization of the present problem, Problem A". Maximize J[y] subject to :

1. $\phi^{\bar{\alpha}}(t, y, \dot{y}) = 0$, $\bar{\alpha} = 1, \ldots, m_1$.
2. $\phi^{\bar{\bar{\alpha}}}(t, y, \dot{y}) \geq 0$, $\bar{\bar{\alpha}} = m_1 + 1, \ldots, m$.
3. $h^{\bar{\beta}}(t_o, t_1, y_o, y_1) = 0$, $\bar{\beta} = 1, \ldots, p_1$.
4. $h^{\bar{\bar{\beta}}}(t_o, t_1, y_o, y_1) \geq 0$, $\bar{\bar{\beta}} = 1, \ldots, p$.

Let $\bar{\phi}$ and $\bar{\bar{\phi}}$ be vectors whose components are $\phi^{\bar{\alpha}}$ and $\phi^{\bar{\bar{\alpha}}}$ respectively and similarly define $\bar{h}$ and $\bar{\bar{h}}$ to be vectors whose components are $h^{\bar{\beta}}$ and $h^{\bar{\bar{\beta}}}$. Let $\phi = \bar{\phi} + \bar{\bar{\phi}}$ and $h = \bar{h} + \bar{\bar{h}}$. Define the set $A = \{1, \ldots, m_1\} \cup \{m_1 + 1 \leq \bar{\bar{\alpha}} \leq m : \omega^{\alpha}(t, \hat{y}, \dot{\hat{y}}) = 0$, for a given $\hat{\Gamma}\}$, and the set $B = \{1, \ldots, p_1\} \cup \{m_1 + 1 \leq \bar{\bar{\beta}} \leq m : h^{\bar{\bar{\beta}}}(\hat{t}_o, \hat{t}_1, \hat{y}_o, \hat{y}_1) =$ $= 0$, for a given $\hat{\Gamma}\}$, where $\hat{\Gamma} = \hat{t}_o, \hat{t}_1, \hat{y}_o, \hat{y}_1$. The sets A and B are indices of constraints that are equality constraints or that are effective at $\hat{\Gamma}$. Let a = the number of elements in A and let b be the number of elements in B.

In section 1 we discuss necessary conditions and in section 2 we discuss sufficient conditions.

1. Necessary conditions.

Theorem 1. <u>If 1) L^o, ϕ and h are of class C^2. 2) $\hat{\Gamma}$ is a solution to problem A". 3) The matrices $[\hat{\phi}^{\alpha}_{\dot{y}}]$, $\alpha \in A$ and $[h^{\beta}_{t_o}, h^{\beta}_{t_1}, h^{\beta}_{y_o}, h^{\beta}_{y_1}]$, $\beta \in B$ have ranks a and b respectively. Then there exist constants $(\lambda_o, \bar{\gamma}. \bar{\bar{\gamma}}) \neq 0$, $\lambda_o \geq 0$ and functions $(\bar{\lambda}, \bar{\bar{\lambda}})$ such that $(\lambda_o, \bar{\lambda}, \bar{\bar{\lambda}}) \neq 0$ at any $t \in [\bar{t}_o, \bar{t}_1]$ such that:</u>

(1.1) (a) $\bar{\bar{\lambda}} \geq 0$ for each of its components, $\lambda^{\bar{\bar{\alpha}}}$, we have

(b) $\lambda^{\bar{\bar{\alpha}}}\hat{\phi}^{\bar{\bar{\alpha}}} = 0$

(1.2) (a) $\bar{\bar{\gamma}} \geq 0$, and for each of its components, $\gamma^{\bar{\bar{\beta}}}$, we have:

(b) $\gamma^{\bar{\bar{\beta}}}\hat{h}^{\bar{\bar{\beta}}} = 0$.

(1.3) Euler-Lagrange-Valentine equations:

$\frac{d}{dt}\hat{F}^2_{\dot{y}} = \hat{F}^2_y$, where $F^2 = \lambda_o L^o + \overline{\lambda\phi} + \bar{\bar{\lambda}}\bar{\bar{\phi}}$.

(1.4) $\frac{d}{dt}(\hat{F}^2 - y\hat{F}^2_{\dot{y}}) = \frac{\partial}{\partial t}\hat{F}^2$

(1.5) Transversality conditions:

i) $\hat{G}^2_{t_o} - \hat{\dot{y}}_o\hat{G}^2_{y_o} - \hat{F}^2\Big|_{t = \hat{t}_o} = 0$

ii) $\hat{G}^2_{t_1} + \hat{\dot{y}}_1\hat{G}^2_{y_1} - \hat{F}^2\Big|_{t = \hat{t}_1} = 0$

iii) $\hat{G}^2_{y_o} - \hat{F}^2_{\dot{y}}\Big|_{t = \hat{t}_o} = 0$

iv) $\hat{G}^2_{y_1} + \hat{F}^2_{\dot{y}}\Big|_{t = \hat{t}_1} = 0$, where $G^2 = \lambda_o h^o + \overline{\gamma h} + \bar{\bar{\gamma}}\bar{\bar{h}}$.

(2) Weierstrass. $E^2(t, \hat{y}, \hat{\dot{y}}, p, \lambda_o, \overline{\lambda}, \bar{\bar{\lambda}}) \leq \bar{\bar{\lambda}}\bar{\bar{\phi}}(t, y, p)$, for every $(\hat{y}, p)$ satisfying the constraints 1. and 2., where $E^2 = F^2(t, y, p, \lambda_o, \lambda) - F^2(t, y, \dot{y}, \lambda_o, \lambda) - (p - \dot{y})\, F^2_{\dot{y}}(t, y, \dot{y}, \lambda_o, \lambda)$.

(3) Clebsch. $\pi\hat{F}^2_{\dot{y}\dot{y}}\pi^* \leq 0$ for all π with $\hat{\phi}^{\alpha}_{\dot{y}}\pi^* = 0$, $\alpha \in A$.

(4) Jacobi-Mayer-Bliss: If $\hat{\Gamma}$ is normal then λ_o, in 1., 2., and 3., is 1 and:
$J^2_2 = d^2G(\tau_o, \tau_1, \eta_o, \eta_1) + [(\hat{F}^2_t - \dot{y}\hat{F}^2_y)(\tau_r)^2 + 2\hat{F}^2_y\eta(t_r)\tau_r]^{r=1}_{r=0} + \int_{t_o}^{t_1}[\eta\hat{F}^2_{yy}\eta^* + 2\eta\hat{F}^2_{y\dot{y}}\eta^* + \dot{\eta}\hat{F}^2_{\dot{y}\dot{y}}\dot{\eta}^*]dt \leq 0$ for p satisfying $\Phi^{\alpha} = 0$ and $\hat{H}^{\beta} = 0$, $\alpha \in A$, $\beta \in B$.

Outline of the proof: We shall now outline the proof of the above theorem. Let us introduce two auxiliary vectors of variables $z(t)$ and $\omega(t)$ with $m - m_1$ and $p - p_1$ component respectively. We may rewrite the constraints 2. as :

(2') $\phi^{\bar{\bar{\alpha}}} - (\dot{z}^{\bar{\bar{\alpha}}})^2 = 0$, $z^{\bar{\bar{\alpha}}}(t_o) = 0$, $z^{\bar{\bar{\alpha}}}(t_1)$ free and the constraints (4) as:

(4') $h^{\bar{\bar{\beta}}} - (\omega^{\bar{\bar{\beta}}}(t_1))^2 = 0$, $\omega^{\bar{\bar{\beta}}}(t_o)$ free.

Now our problem is to maximize J subject to (1), (2'), (3) and (4') which is a problem of type A. Define $F = \lambda_o L^o + \sum_{\bar{\alpha}}\lambda^{\bar{\alpha}}\phi^{\bar{\alpha}} + \sum_{\bar{\bar{\alpha}}}\lambda^{\bar{\bar{\alpha}}}(\phi^{\bar{\bar{\alpha}}} - (z^{\bar{\bar{\alpha}}})^2)$ and $G = \lambda_o h^o + \sum_{\bar{\beta}}\gamma^{\bar{\beta}}h^{\bar{\beta}} + \sum_{\bar{\bar{\beta}}}\gamma^{\bar{\bar{\beta}}}(h^{\bar{\bar{\beta}}} - (\omega^{\bar{\bar{\beta}}}(t_1)^2)$

By the Euler-Lagrange equations for problem A, condition (1), theorem 7, chapter 5, we have:

(1.1) $\frac{d}{dt}\hat{F}_{\dot{y}} = \hat{F}_y$.

(1.2) $\frac{d}{dt}\hat{F}_{\dot{z}} = \hat{F}_z$.

But, $F_{\dot{z}^{\bar{\bar{\alpha}}}} = -2\lambda^{\bar{\bar{\alpha}}}\hat{\dot{z}}^{\bar{\bar{\alpha}}}$ and $\hat{F}_{z^{\bar{\bar{\alpha}}}} = 0$. Thus by (1.2),

(2) $\mu^{\bar{\bar{\alpha}}}\hat{\dot{z}}^{\bar{\bar{\alpha}}}$ = constant.

By transversality conditions for problem A, conditions (3) of theorem 7, chapter 5, we have:

(3.1) $G_{t_o} + \hat{\dot{y}}_o\hat{G}_{y_o} + \hat{\dot{\omega}}_o\hat{G}_{\omega_o} - \hat{F}\Big|_{t = t_o} = 0$

(3.2) $G_{t_1} + \hat{\dot{y}}_1\hat{G}_{y_1} + \hat{\dot{z}}_1\hat{G}_{z_1} + \hat{\dot{\omega}}_1\hat{G}_{\omega_1} - \hat{F}\Big|_{t = t_1} = 0$

(3.3.1) $\hat{G}_{y_o} - \hat{F}_{\dot{y}}\Big|_{t = \hat{t}_o} = 0.$

(3.3.2) $\hat{G}_{\omega_o} - \hat{F}_{\dot{\omega}}\Big|_{t = \hat{t}_o} = 0.$

(3.4.1) $\hat{G}_{y_1} + \hat{F}_{\dot{y}}\Big|_{t = \hat{t}_1} = 0.$

(3.4.2) $\hat{G}_{z_1} + \hat{F}_{\dot{z}}\Big|_{t = \hat{t}_1} = 0.$

(3.4.3) $\hat{G}_{\omega_1} + \hat{F}_{\dot{\omega}}\Big|_{t = \hat{t}_1} = 0.$

By definition of F and G we have: $\hat{G}_{\omega_o} = 0$, $\hat{F}_{\dot{\omega}} = 0$, $\hat{G}_{z_1} = 0$, $\hat{F}_{\dot{z}^{\bar{\bar{\alpha}}}} = -2\mu^{\bar{\bar{\alpha}}}\dot{z}^{\bar{\bar{\alpha}}}$,

$\hat{G}_{\omega_1^{\bar{\bar{\beta}}}} = -2\gamma^{\bar{\bar{\beta}}}\hat{\omega}^{\bar{\bar{\beta}}}$ and $\hat{F}_{\dot{\omega}} = 0$. Thus we write (3) as:

(4.1) $\hat{G}_{t_o} + \hat{\dot{y}}_o\hat{G}_{y_o} - \hat{F}\Big|_{t = \hat{t}_o} = 0$

(4.2)' $\hat{G}_{t_1} + \hat{\dot{y}}_1\hat{G}_{y_1} + \hat{\dot{\omega}}_1\hat{G}_{\omega_1} + \hat{F}\Big|_{t = \hat{t}_1} = 0$

(4.3) $\hat{G}_{y_o} - \hat{F}_{\dot{y}}\Big|_{t = \hat{t}_o} = 0$

(4.4.1) $G_{y_1} + \hat{F}_{\dot{y}}\Big|_{t = t_1} = 0$

(4.4.2) $\hat{F}_{\dot{z}^{\bar{\bar{\alpha}}}}\Big|_{t = t_1} = -2\lambda^{\bar{\bar{\alpha}}}(t_1)\dot{z}^{\bar{\bar{\alpha}}}(t_1) = 0$

(4.4.3) $\hat{G}_{\omega_1^{\bar{\bar{\beta}}}} = -2\gamma^{\bar{\bar{\beta}}}\hat{\omega}_1^{\bar{\bar{\beta}}} = 0.$

By (4.4.3), (4.2)' becomes:

(4.2) $\hat{G}_{t_1} + \hat{\dot{y}}_1\hat{G}_{y_1} + \hat{F}\Big|_{t = t_1} = 0.$

i. Proof of 1: By (4.4.2) and (2) we have: $\mu^{\bar{\bar{\alpha}}}\dot{\hat{z}}^{\bar{\bar{\alpha}}} = 0$, which establishes (1.1)-b of the theorem, since $\dot{\hat{z}}^{\bar{\bar{\alpha}}} \neq 0$ iff: $\hat{\phi}^{\bar{\bar{\alpha}}} > 0$. By (4.4.3) we have: $\gamma^{\bar{\bar{\beta}}}\hat{h}^{\bar{\bar{\beta}}} = 0$ which is condition (1.2)-b of the theorem.

Now take $\lambda^{\bar{\alpha}}$ to be the components of $\bar{\lambda}$ of the theorem and $\lambda^{\bar{\bar{\alpha}}}$ as the components of $\bar{\bar{\lambda}}$. Also take $\gamma^{\bar{\beta}}$ and $\gamma^{\bar{\bar{\beta}}}$ as the components of $\bar{\gamma}$ and $\bar{\bar{\gamma}}$. It is then obvious that the "derivatives", with respect to y, $\dot{y}$, t_o, t_1, y_o and y_1 of F and G are equal to the derivatives of F^2 and G^2 along $\hat{\Gamma}$. This establishes (1.3), (1.4) and (1.5) of the theorem.

We now outline the proof of parts (1.1)-a. and (1.2)-a. of the theorem. Fix $\tilde{t} \in [\hat{t}_o, \hat{t}]$. If $\bar{\bar{\alpha}} \notin A$, then $\lambda^{\bar{\bar{\alpha}}} = 0$, by (1.1)-b. We now show that $\lambda^{\bar{\bar{\alpha}}} \geq 0$ for $\bar{\bar{\alpha}} \in A$. At $\tilde{t}$ we may, without loss of generality, ignore, locally, the constraints $\phi^{\bar{\bar{\alpha}}}$ for $\bar{\bar{\alpha}} \notin A$. By the Clebsch condition for this problem of type A we have:

$$\pi_1\hat{F}_{\dot{y}\dot{y}}\pi_1^* + \pi_1\hat{F}_{\dot{y}\dot{z}}\pi_2^* + \pi_2\hat{F}_{\dot{z}\dot{y}}\pi_1^* + \pi_2\hat{F}_{\dot{z}\dot{z}}\pi_2^* \leq 0 \text{ for } \pi_1, \pi_2 \text{ satisfying:}$$

(5.1) $\hat{\phi}^{\bar{\alpha}}\pi_1^* + \hat{\phi}^{\bar{\alpha}}_{\dot{z}}\pi_2^* = \hat{\phi}^{\bar{\alpha}}_{\dot{y}} = 0$

(5.2) $\hat{\phi}^{\alpha}_{\dot{y}}\pi_1^* + \hat{\phi}^{\bar{\bar{\alpha}}}_{\dot{z}}\pi_2^* = 0, \ \alpha \in A$

But $\hat{\phi}^{\bar{\bar{\alpha}}}_{\dot{z}} = -2\dot{\hat{z}}^{\bar{\bar{\alpha}}} = 0, \ \alpha \in A$. Thus (5) does not restrict π_2. Furthermore, $\pi_1 = 0$ satisfies (5). Also $\hat{F}_{\dot{y}\dot{z}} = 0$ and $\hat{F}_{\dot{z}\dot{y}} = 0$. Now take $\pi_1 = 0$ and $\pi_2 = 0$ except for one component $\bar{\bar{\alpha}}$. The inequality still holds and $F_{\dot{z}^{\bar{\bar{\alpha}}}\dot{z}^{\bar{\bar{\alpha}}}}(\pi^{\bar{\bar{\alpha}}}) = -2\lambda^{\bar{\bar{\alpha}}}(\pi^{\bar{\bar{\alpha}}})^2 \leq 0$. Hence $\lambda^{\bar{\bar{\alpha}}} \geq 0$ for $\bar{\bar{\alpha}} \in A$.

We now indicate the proof of (1.2)-b. Going back to the outline of the proof of theorem 7, after reducing problem A to a finite dimensional problem. Noting that the first order necessary condition of chapter 3 applies, we conclude that $\gamma^{\bar{\bar{\beta}}} \geq 0$.

ii. Proof of 2: Recalling the definition of $F^{(1)}$ in (i), we have:

(6) $E(t, \hat{y}, \hat{z}, \hat{y}, \hat{z}, p, \tilde{p}, \lambda_o, \lambda)^{(2)} = F^2(t, \hat{y}, p) - \sum_{\alpha} \lambda^{\bar{\bar{\alpha}}}(\tilde{p}^{\bar{\bar{\alpha}}})^2 - \hat{F}^2 + \sum_{\bar{\bar{\alpha}}} \lambda^{\bar{\bar{\alpha}}}(\hat{z}^{\bar{\bar{\alpha}}})^2 - (p - y)F^2_{\dot{y}} - (p - z)F_{\dot{z}}.$

---

(1) $F = F^2 - \sum_{\bar{\bar{\alpha}}} \lambda^{\alpha}(z^{\bar{\bar{\alpha}}})^2$

(2) p is the "comparison" vector of z.

The fourth and last terms of the right hand side of (6) are zeros, by i, and we have:

$$E(t, \hat{y}, \hat{z}, \dot{\hat{y}}, \dot{\hat{z}}, p, \tilde{p}, \lambda_o, \lambda) = E^2(t, \hat{y}, \dot{\hat{y}}, p, \lambda_o, \lambda) - {}_{\alpha} \lambda^{\bar{\bar{\alpha}}}(\tilde{p}^{\bar{\bar{\alpha}}})^2.$$

By the Weierstrass condition for problem A, $E \leq 0$ for $(\hat{y}, p, \tilde{p})$ satisfying constraints (1) and (2)', i.e., satisfying (1) and $(\tilde{p}^{\bar{\bar{\alpha}}})^2 = \phi^{\bar{\bar{\alpha}}}$. Thus $E^2(t, \hat{y}, \dot{\hat{y}}, p, \lambda_o, \lambda) - \lambda\phi \leq 0$ as we were to show.

iii. Proof of 3: Condition (3) follows immediately from the Clebsch condition for problem A, arguing as we did towards the end of i. above.

iv. Proof of 4: By the second order necessary condition for problem A we have $\hat{J}_2 \leq 0$ for $\hat{\Phi} = 0$ and $\hat{H} = 0$. But $\hat{J}_2 = d^2\hat{G}^2 - 2 {}_{\bar{\bar{\beta}}} \gamma^{\bar{\bar{\beta}}}(\omega^{\bar{\bar{\beta}}})^2 + [(\hat{F}^2_t - \dot{y}\hat{F}^2_y)(\tau^2_r) + 2\hat{F}^2_y\eta(t_r)\tau_o]\Big|_{r=0}^{r=1} + \int_{\hat{t}_o}^{\hat{t}_1} \eta\hat{F}^2_{yy}\eta^* + 2\eta\hat{F}^2_{y\dot{y}}\dot{\eta}^* + \dot{\eta}\hat{F}^2_{\dot{y}\dot{y}}\dot{\eta}dt - \int_{\hat{t}_o}^{\hat{t}_1} {}_{\bar{\bar{\alpha}}} \mu^{\bar{\bar{\alpha}}}(\zeta^{\bar{\bar{\alpha}}})^2$, where ω and ζ are the increments of w and z, since $F_t + F^2_t$, $F_z = 0$, $F_{\dot{z}z} = 0$. Thus:

$$\hat{J}_2 + \hat{J}_2 - {}_{\bar{\bar{\beta}} \varepsilon B} \gamma^{\bar{\bar{\beta}}}(\omega^{\bar{\bar{\beta}}})^2 - {}_{\bar{\bar{\alpha}} \varepsilon A} \int_{\hat{t}_o}^{\hat{t}_1} \mu^{\bar{\bar{\alpha}}}(\zeta^{\bar{\bar{\alpha}}})^2,$$

since $\gamma^{\bar{\bar{\beta}}}$ and $\mu^{\bar{\bar{\alpha}}}$ are zeros if the corresponding constraints are ineffective. Thus the choice of "increments" $\omega^{\bar{\bar{\beta}}}$, $\bar{\bar{\beta}} \notin B$ and $\zeta^{\bar{\bar{\alpha}}}$, $\bar{\bar{\alpha}} \notin A$ does not affect $\hat{J}_2$. Writing down $\Phi^{\alpha} = 0$, $\alpha \varepsilon A$ and $H^{\beta} = 0$, $\beta \varepsilon B$ we find that the choice of the remaining $\omega^{\bar{\bar{\beta}}}$'s and $\zeta^{\bar{\bar{\alpha}}}$'s is arbitrary since they have zero coefficients in the expressions. Thus we may chase them to be zeros and $\hat{J}_2 = \hat{J}^2_2$.

Finally we note that rank conditions and normality of problem A characterization are easily verified in view of the corresponding conditions for this problem.

2. Sufficient conditions.

Theorem 2. Let L°, ϕ, and h be of class C^2 and let $\hat{\Gamma}$ satisfy conclusion (1) of theorem 1 of this chapter with $\lambda_o > 0$. Then:

a) If $\phi^{\bar{\alpha}}$ and $h^{\bar{\beta}}$ are linear and L°, $\phi^{\bar{\bar{\alpha}}}$, $h^{\bar{\bar{\beta}}}$ are concave in y and $\dot{y}$, then $\hat{\Gamma}$ is a global solution of problem A".

b) If $\hat{J}^2_2 < 0$ for $p \neq 0$ with $\hat{\Phi}^{\alpha} = 0$, $\alpha \varepsilon A$ and $\hat{H}^{\beta} = 0$, $\beta \varepsilon B$, then $\hat{\Gamma}$ is a weak solution to problem A".

Part a) follows from the first order sufficient condition for problem A, since our hypothesis implies that the "path" and "end point" Lagrangians are concave.

Part b follows from Penmissi's theorem [41] since our set of variations p with Φ^{α} = = 0, H^{β} = 0, $\alpha \in A$, $\beta \in B$ includes his.

CHAPTER 7

EXTENSIONS AND APPLICATIONS

In this chapter we note some extensions of the theorems of chapter 6 and some applications. We first state a problem in optimal control and characterize its solutions as applications of chapter 6. The optimal control problem with scaler criterion is presented in section 1. In section 2 we present extensions of the control problem to: a) problems with time lags, b) problems with bounded state variables and, c) problems with finite vector criteria. For these problems we discuss only the first order necessary conditions and the Weierstrass conditions. In sections 3 and 4 we present some economic applications.

1. An optimal control problem.

First we state an optimal control problem with scaler criterion and characterize it. Next we discuss normality. Then we discuss two special problems: a) when the controls are restricted to be non-negative and b) when the criterion is a function of integrals.

1.1 Statement of the problem.

Suppose a dynamical system is given by:

(1) $\dot{x} = f(t, x, u)$,

where x is an n-vector function and u is an R vector function. x will be referred to as the state variable vector and u will be referred to as the control vector. Our problem is to maximize:

$$J[u] = g^o(t_o, t_1, x_o, x_1) + \int_{t_o}^{t_1} f^o(t, x, u)\, dt,$$

where $x_o = x(t_o)$, $x_1 = x(t_1)$, subject to (1) and subject to:

(2.1) $K^{\bar{\alpha}}(t, x, u) = 0, \qquad \bar{\alpha} = 1, \ldots, m_1$

(2.2) $K^{\bar{\bar{\alpha}}}(t, x, u) \geq 0, \qquad \bar{\bar{\alpha}} = m_1 + 1, \ldots, m$

(3.1) $g^{\bar{\beta}}(t_o, t_1, x_o, x_1) = 0, \qquad \bar{\beta} = 1, \ldots, p_1$

(3.2) $g^{\bar{\bar{\beta}}}(t_o, t_1, x_o. x_1) \geq 0, \qquad \bar{\bar{\beta}} = 1, \ldots, p_2.$

<u>Note</u>: We have suppressed the arguments of x(t) and u(t).

Let A be the set of indices α corresponding to either equality constraints

(2.1) or to effective inequality constraints* (2.2) at some arc $\hat{\Delta} = (\hat{\Delta}_1; \hat{\Delta}_2) =$ $= (\hat{t}_o, \hat{t}_1, \hat{x}_o, \hat{x}_1; \hat{x}, \hat{u})$. And let B be the set of indices β corresponding to equality constraints (3.1) or to effective inequality constraints (3.2).

As before, we say that an arc is normal if it is the first order necessary condition with a unique vector of multipliers with $\lambda_o = 1$. A necessary and sufficient condition for normality of a solution to the control problem will be stated in section 1.3. Finally, denote the variation of x(t) by $\xi(t)$, the variation of u(t) by ν and let τ_o and τ_1 be as before. Denote the variation vector by $S = (\tau_o, \tau_1, \xi_o, \xi_1, \xi, \nu)$.

1.2 Characterization of the solution.

Theorem 1. Necessary conditions (Pontriagin-Hestenes-Berkovitz). If: 1) $\hat{\Delta} = (\hat{t}_o, \hat{t}_1, \hat{x}_o, \hat{x}_1; \hat{x}, \hat{u})$ be a solution to the control problem. 2) The functions g^o, g, f^o, f and K are in class C^2. 3) The matrices $[\hat{K}^\alpha_u]$, $\alpha \in A$ and $[\hat{g}^\beta_{\Delta_1}]$, $\beta \in B$ have maximal ranks. Then there exist constants $(\lambda_o, \overline{\gamma}, \bar{\gamma})$ and time functions $(\lambda(t), \overline{\mu}(t), \bar{\bar{\mu}}(t))$ with $(\lambda_o, \overline{\gamma}, \bar{\bar{\gamma}}) \neq 0$ and $(\lambda_o, \lambda, \mu) \neq 0$ such that:

1.i. $\bar{\bar{\mu}} \geq 0$ and $\bar{\bar{\mu}}\hat{\bar{\bar{K}}} = 0$.

1.ii. $\bar{\bar{\gamma}} \geq 0$ and $\bar{\bar{\gamma}}\hat{\bar{\bar{g}}} = 0$.

1.iii. (Euler-Langrange):

a) $\dot{\lambda} = -\hat{\mathcal{H}}_x$, where $\mathcal{H} = \lambda_o f^o + \lambda f + \overline{\mu K} + \bar{\bar{\mu}}\bar{\bar{K}} = \mathcal{H}(t, \lambda_o, \lambda, \mu, x, u)$.

b) $\hat{\mathcal{H}}_u = 0$

1.iv. $\frac{d}{dt}\hat{\mathcal{H}} = \hat{\mathcal{H}}_t$.

2. (Weierstrass condition-Pontriagin maximum principle*): With $\lambda(t)$ defined in 1.ii we have:

$$\lambda_o f^o(t, \hat{x}, \hat{u}) + \lambda f(t, \hat{x}, \hat{u}) \geq \lambda_o f^o(t, \hat{x}, u) + \lambda f(t, \hat{x}, u)$$

for $\hat{x}$, u satisfying constraints (1) and (2).

(*) An inequality constraint is effective at a point if it holds as an equation at that point.

(*) See Pontriagin, etc. [42].

3. (Transversality conditions):

i) $\hat{G}_{t_o} - \hat{\mathcal{H}}\Big|_{t = \hat{t}_o} = 0$

ii) $\hat{G}_{t_1} + \hat{\mathcal{H}}\Big|_{t = \hat{t}_1} = 0$

iii) $\hat{G}_{x_o} - \lambda(\hat{t}_o) = 0$

iv) $\hat{G}_{x_1} + \lambda(\hat{t}_1) = 0$

where $G = \lambda_o g^o + \lambda g$.

4. (Clebsch-Valentine-Berkovitz) $\pi \hat{\mathcal{H}}_{uu}\pi^* \leq 0$ for π with:

$$\hat{K}^{\alpha}_u \pi^* = 0, \ \alpha \in A.$$

5. (Jacobi-Mayer-Bliss) If, in addition to our assumptions, Γ is normal, then:

6. $Q(\hat{\Delta}, \lambda_o, \lambda, \mu, \gamma; \delta) = d^2G(\hat{\Delta}_1; \tau_o, \tau_1, \xi_o, \xi_1) + [(\hat{\mathcal{H}}_t + \dot{\lambda}f)\tau_r^2 + 2\dot{\lambda}\xi(t_r)\tau_r]\Big|_{r=0}^{=1}$

$+ \int_{t_o}^{\hat{t}_1} [\xi\hat{\mathcal{H}}_{xx}\xi^* + 2\xi\hat{\mathcal{H}}_{xu}\nu^* + \nu\hat{\mathcal{H}}_{uu}\nu^*]dt \leq 0$ for δ satisfying:

7.1. $\dot{\xi} = \hat{f}_x\xi^* + \hat{f}_u\nu^*$

7.2. $\hat{K}^{\alpha}_x\xi + \hat{K}^{\alpha}_u\nu = 0, \ \alpha \in A$

7.3. $(\hat{g}^{\beta}_{t_o} + \hat{g}^{\beta}_{x_o}\hat{f}\Big|_{t = t_1})\tau_o + (\hat{g}^{\beta}_{t_1} + \hat{g}^{\beta}_{x_1}\hat{f}\Big|_{t = t_1})\tau_1 + \hat{g}^{\beta}_{x_o}\xi_o + \hat{g}^{\beta}_{x_1}\xi_1 = 0, \ \beta \in B.$

Outline of the proof: Introduce the variables y as follows.

8. $\dot{s} = u.$

We may rewrite our problem as:

$$\text{Maximize } J = g(t_o, t_1, x_o, x_1 + \int_{t_o}^{t_1} f^o(t, x, \dot{s}) \text{ s.t}_o$$

(1)' $f(t, x, \dot{s}) - \dot{x} = 0$

(2.1)' $\overline{K}^{\alpha}(t, x, \dot{s}) = 0$

(2.2)' $\overline{\overline{K}}^{\alpha}(t, x, \dot{s}) \geq 0,$

and to (3.1) and (3.2)

This now is a problem of type A". $\hat{\Gamma} = (\hat{t}_o, \hat{t}_1, \hat{x}_o, \hat{x}_o, \hat{y}_o, \hat{y}_1, \hat{x}, \hat{s})$ is a solution to this problem. Define $F^2 = \lambda_o f^o + \lambda(f - \dot{x}) + \overline{\mu K} + \overline{\overline{\mu K}}$, and $G^2 = \lambda_o g^o + \overline{\gamma g} + \overline{\overline{\gamma g}}$. Since g^o, g is independent of s_o and s_1 we have G^2 is identical to G and we obtain condition 1.ii as an immediate result of 1.2 of theorem 1 of chapter 6. That theorem applies as could easily be verified. We further note that $F^2 = \mathcal{H} - \lambda\dot{x}$, and that results in 1.i which follows from 1.1 of theorem 1 of chapter 6, by taking $(\lambda, \bar{\mu})$ $\bar{\lambda}$ and $\bar{\bar{\mu}}$ to $\bar{\bar{\lambda}}$. Now by the Euler-Lagrange equations for problem A" we have: $\frac{d}{dt}\hat{F}^2_{\dot{x}} = \hat{F}^2_x$ and $\frac{d}{dt}\hat{F}^2_{\dot{s}} = \hat{F}^2_s$. But $\hat{F}^2_{\dot{x}} = -\lambda$ and $\hat{F}^2_x = \hat{\mathcal{H}}_x$. Thus $\dot{\lambda} = -\hat{H}_x$ which is 1.iii-a of the present theorem. Also $\hat{F}^2_{\dot{s}} = \hat{\mathcal{H}}_{\dot{s}} = \hat{\mathcal{H}}_u$ and $\hat{F}_s = 0$. Thus $\frac{d}{dt}\hat{\mathcal{H}}_u = 0$, i.e., $\hat{\mathcal{H}}_u =$ constant. By the transversality condition, 1.5(iii), theorem 1 of chapter 6, $\hat{F}_{\dot{s}} = 0$. Thus $\hat{\mathcal{H}}_u = 0$ which is our condition 1.iii-b. $\hat{F}^2_t = \hat{\mathcal{H}}_t$, by definition of F^2 and $\mathcal{H}$. Also $\hat{F}^2_{\dot{s}} = \hat{F}^2_u = 0$ as we have just seen and $\hat{F}^2_{\dot{x}} = -\lambda$. Thus, letting y in problem A" equal (x, s), we have from condition (1.4) of theorem 1, chapter 6:

$$\frac{d}{dt}(\hat{F}^2 - \dot{\hat{y}}\hat{F}^2_{\dot{y}}) = \frac{d}{dt}(\hat{F}^2 + \lambda\dot{x}) = \frac{d}{dt}\hat{\mathcal{H}} = \hat{F}^2_t = \hat{\mathcal{H}}_t,$$

which is our condition 1.iv.

To prove 2, we write the Weierstrass E function, E^2, for the present problem. Let $p = (p_1, p_2)$ be the "test function" for our problem, corresponding to $(\hat{x}, \dot{y}) = (\hat{x}, \hat{u})$. Since the constraints must be satisfied for the inequality, 2 of theorem 1, chapter 6, to hold we have $p_1 = f(t, x, u)$. Writing the inequality and noting the relation between F^2 and $\mathcal{H}$ we have:

$$E^2 = \mathcal{H}(t, \hat{x}, u, \lambda_o, \lambda, \mu) - \lambda f(t, \hat{x}, u) - \mathcal{H}(t, \hat{x}, \hat{u}, \lambda_o, \lambda, \mu) + \\ + \lambda f(t, \hat{x}, \hat{u}) - (f(t, \hat{x}, u) - \dot{\hat{x}})\hat{F}^2_{\dot{x}} - (u - \hat{u})\hat{F}^2_u \le \bar{\bar{\mu}}\bar{\bar{K}}.$$

But $\hat{F}^2_{\dot{x}} = -\lambda$ and $\hat{F}^2_u = 0$. Thus we have:

$$\mathcal{H}(t, \hat{x}, u, \lambda_o, \lambda, \mu) - \mathcal{H}(t, \hat{x}, \hat{u}, \lambda_o, \lambda, \mu) - \lambda f(t, \hat{x}, u) + \\ + \lambda f(t, \hat{x}, \hat{u}) - f(t, \hat{x}, u) + f(t, \hat{x}, \hat{u}) - \bar{\bar{\mu}}\bar{\bar{K}} \le 0,$$

But $\mathcal{H}(t, \hat{x}, \hat{u}, \lambda_o, \lambda, \mu) = \lambda_o f^o(t, \hat{x}, \hat{u}) + \lambda f(t, \hat{x}, \hat{u})$, by 1.i and since $\bar{K}(t, \hat{x}, \hat{u}) = 0$. Also $\mathcal{H}(t, \hat{x}, u) - \bar{\bar{\mu}}\bar{\bar{K}} = \lambda_o f^o(t, \hat{x}, u) + \lambda f(t, \hat{x}, u)$ in case constraints (2.1) are satisfied. Combining this with our last inequality we get 2. of the theorem

To prove 3., we write the transversality conditions for the present problem recalling that $G^2 = G$ and substituting $F^2 = \mathcal{H} - \lambda\dot{x}$. By (1.5) of theorem 1 of chapter 6 we have:

9. $\hat{G}_{t_o} - \dot{\hat{x}}_o\hat{G}_{x_o} - \hat{\mathcal{H}}\Big|_{t = t_o} + \lambda(t_o)\dot{\hat{x}}_o = 0$

10. $\hat{G}_{t_1} + \dot{\hat{x}}_1 G_{x_1} + \hat{\mathcal{H}}\Big|_{t = t_1} + \lambda(t_1)\dot{\hat{x}}_1 = 0$

11. $\hat{G}_{x_o} - \lambda(t_o) = 0$

12. $\hat{G}_{x_1} + \lambda(t_1) = 0.$

Relations 11. and 12. establish iii and iv of 3. in the theorem. Substituting from 11. and 12. for $\hat{G}_{x_o}$ and $\hat{G}_{x_1}$ in 9. and 10. we have:

13. $\hat{G}_{t_o} - \dot{\hat{x}}_o\lambda(t_o) - \hat{\mathcal{H}}\Big|_{t = t_o} + \lambda(t_o)\dot{\hat{x}}_o = \hat{G}_{t_o} - \hat{\mathcal{H}}\Big|_{t = t_o} = 0$

14. $\hat{G}_{t_1} + \dot{\hat{x}}_1\lambda(t_1) + \hat{\mathcal{H}}\Big|_{t = t_1} + \lambda(t_1)\dot{\hat{x}}_1 = \hat{G}_{t_1} + \hat{\mathcal{H}}\Big|_{t = t_1} = 0.$

But 13. and 14. are i and ii of 3. Thus 3. is "established".

4. follows from 3. of theorem 1, chapter 6, by noting that $\hat{F}^2_{\dot{x}\dot{x}} = 0$ and that $\hat{F}^2_{uu} = \hat{\mathcal{H}}_{uu}$. 5. follows from 4. of theorem 1, chapter 6 by noting that:

i) $d^2G^2 = d^2G$

ii) $F^2_t = \hat{\mathcal{H}}_t$

iii) $\dot{\hat{x}}\hat{F}^2_x = -\dot{\lambda}f(t, \hat{x}, \hat{u}), u\hat{F}^2_z = 0$

iv) $\hat{F}^2_{xx} = \hat{\mathcal{H}}_{xx}, \hat{F}^2_{xz} = 0, \hat{F}^2_{zx} = 0, \hat{F}^2_{zz} = 0.$

v) $\hat{F}^2_{x\dot{x}} = 0, \hat{F}^2_{xu} = 0, \hat{F}^2_{z\dot{x}} = 0, \hat{F}^2_{zu} = 0$

vi) $\hat{F}^2_{\dot{x}u} = 0, \hat{F}^2_{\dot{x}\dot{x}} = 0, \hat{F}^2_{u\dot{x}} = 0, \hat{F}^2_{uu} = \mathcal{H}_{uu}.$

This completes an outline of the proof of theorem 1.

Theorem 2. Sufficient conditions (Mangasarian, Pennissi): Suppose: 1) g^o, g, f^o, f and K are of class C^2. 2) $\hat{\Delta}$ satisfies conditions 1. and 3. of theorem 1 with $\lambda_o > 0$. Then:

a) If f, $\overline{g}$ and $\overline{K}$ are linear and if f^o, $\overline{\overline{K}}$ and $\overline{\overline{g}}$ are concave, then $\hat{\Delta}$ is a global solution of the contro problem.

b) If $Q < 0$ for $\delta \neq 0$ satisfying equations (7) of this section, where Q is defined in 6. of theorem 1, then $\hat{\Delta}$ is a weak local solution of the control problem.

Theorem 2 follows in a straightforward manner from theorem 2 of chapter 6 after performing the necessary transformations as we did in theorem 1 of this section.

1.3 Normality

Here we state a necessary and sufficient condition for normality in the control problem. However, the condition is hard to verify and, in many applications, one may verify that $\lambda_o > 0$ by way of contradiction. This, of course, would not be a proof of normality since the multiplier vector may not be unique.

Remark: The rank condition is necessary and sufficient for Δ to be normal.

Definition: The rank condition for optimal control: We say that the rank condition for optimal control is satisfied if the matrix $[(\hat{g}^{\beta}_{t_o} + \hat{g}^{\beta}_{x_o}\dot{\hat{x}}_o)\tau^s_o + (\hat{g}^{\beta}_{t_1} + \hat{g}^{\beta}_{x_1}x_1)\tau^{\sigma} + + \hat{g}^{\beta}_{x_o}\xi^{\sigma}_o + \hat{g}^{\beta}_{x_1}\xi^{\sigma}_1]$, $\beta \in B$, where τ^{σ}_o, τ^{σ}_1, ξ^{σ}_o, ξ^{σ}_1 are the end points of a variation $(\xi^{\sigma}, \nu^{\sigma})$ with respect to a vector of parameters with as many components as the number of elements of B, has rank equal to the number of elements of B for e ery such admissible variation. An admissible variation is, as before, a variation such that the differentials of equality constraints and effective inequality constraints, with $(\xi^{\sigma}, \nu^{\sigma})$ as increments, is zero.

1.4 Two special problems.

(a) If, in addition to the constraints (1)-(3) we have: $u \geq 0$, $x_o \geq 0$, $x_1 \geq 0$, then the condition 1.iii-b of theorem 1 becomes $\hat{\mathcal{H}}_u \leq 0$, $\hat{\mathcal{H}}_{u_r}\hat{u}_r = 0$, and conditions 3.iii and 3.iv of the theorem become: $\hat{G}_{x_o} \leq \lambda(t_o)$, $\hat{x}^i_o\hat{G}_{\hat{x}^i_o} = \lambda^i(t_o)\hat{x}^i_o$ and $\hat{G}_{x_1} \leq \lambda(t_1)$, $\hat{x}_1\hat{G}_{x^i_1} = \hat{x}^i_1\lambda(t_1)$. This could be easily seen by considering the non-negativity constraints as additional inequality constraints of the form (2.2) and (3.2)

(b) Suppose our objective function is of the form: $g^o + W[\int_{t_o}^{t_1} C(t, x, u)dt]$, where C is an n-vector valued function. Then the Hamiltonian for the problem becomes: $\mathcal{H}' = \lambda_o W_C C + \lambda f$ and G remains unchanged. This could be seen by introducing the variables θ as follows:

15. $\dot{\theta} = C(t, x, u)$, $\theta(t_o)$ and $\theta(t_1)$ free.

The problem then becomes: max. $g^o + W(\theta_1 - \theta_o)$ subject to 1. - 3. and to 15.. The multipliers corresponding to 15. are constants and their values are determined from the transversality conditions of theorem 1. This problem was first formulated by Brady [12].

2. Extensions.

2.1 Problem with bounded state variables.

If we have, in addition to 1. - 3., constraints of the form:

16. $M^j(t, x) \geq 0$, $j = 1, \ldots, s$,

then the rank condition in the first order necessary theorem could not be satisfied. Here we state a modification to the Euler-Lagrange equations and the Pontriagin maximum principle of that theorem, due to Russak [45]. Add to the assumptions of theorem 1 the assumption : M is of class C^2. Introduce the functions $N(t, x, u) = M_t + M_x f(t, x, u)$ and the multipliers $q(t)$. Define $\mathcal{H}'' = \mathcal{H} + qN$. Then, if $\hat{\Delta}''$ is a solution to this problem, conditions 1., 2., and 3. of the theorem hold for $\mathcal{H}''$ with G unchanged. Furthermore:

(i) $p(t) \geq 0$,

(ii) $\dot{p}(t) \leq 0$,

(iii) $\dot{p}(t) = 0$ on intervals with $M(t, \hat{x}) > 0$ and

(iv) $p(t_r)M(\hat{t}_r, \hat{x}_r) = 0$, $r = 0, 1$.

2.2 Problems with time lags.

We discuss a control problem with time lags in both state and control variables with equality and inequality constraints. We present a conjecture about the Euler-Lagrange equations and the Pontriagin maximum principle. In the absence of inequality constraints, the conjecture was proved by Halanay* [24]. Consider a system given by:

17. $\dot{x}(t) = F(t, x, x^{-1}, \ldots, x^{-k}, u, u^{-1}, \ldots, u^{-k})$,

where $x = x(t)$ is an n-vector valued function, $u = u(t)$ is a R-valued function, $x^{-s} = x(t - \tau^s)$ and $u^{-s} = u(t - \tau^s)$ with $s = 1, \ldots, k$, τ^s fixed and positive. Let t_o and t_1 be fixed and let $x(t)$ be a given function on the interval $[t_o - \max_\beta \tau^s, t_o]$

Our problem is to maximize the functional:

---

* Halanay description of the control systems is much more general than the one presented here.

$$g^o(x_1) + \int_{t_o}^{t_1} f^o(t, x, x^{-1}, \ldots, x^{-k}, u, u^{-1}, \ldots, u^{-k}),$$

subject to:

17.1. $K^{\bar{\alpha}}(t, x, x^{-1}, \ldots, x^{-k}, u, u^{-1}, \ldots, u^{-k}) = 0, \bar{\alpha} = 1, \ldots, m_1,$

17.2. $K^{\bar{\bar{\alpha}}}\ t, x, x^{-1}, \ldots, x^{-k}, u, u^{-1}, \ldots, u^{-k}) \geq 0, \bar{\bar{\alpha}} = m_1 + 1, \ldots, m.$

18.1. $g^{\bar{\beta}}(x_1) = 0, \bar{\beta} = 1, \ldots, p_1.$

18.2. $g^{\bar{\bar{\beta}}}(x_1) \geq 0, \bar{\bar{\beta}} = p_1 + 1, \ldots, p.$

Let η be a time function, by η^{-s} we denote $\eta(t - \tau^s)$, by η^r we denote $\eta(t + \tau^r)$ and by η^{r-s} we denote $\eta(t - \tau^s + \tau^r)$. Let $\lambda(t)$, $\bar{\mu}(t)$ and $\bar{\bar{\mu}}(t)$ be time functions with the dimensions f, $\bar{K}$ and $\bar{\bar{K}}$ and let $\lambda_o > 0$ be a constant scaler. Let $\bar{\gamma}$ and $\bar{\bar{\gamma}}$ be constant vectors with the dimensions of $\bar{g}$ and $\bar{\bar{g}}$. Define:

$$\mathcal{H}^o = \lambda_o f^o + \lambda f + \bar{\mu}\bar{K} + \bar{\bar{\mu}}\bar{\bar{K}} = \lambda_o f^o + \lambda f + \mu k.$$

$$\mathcal{H}^1 = \lambda_o f^o(t + \tau^1, x^1, x^{1-2}, \ldots, x^{1-k}, u^1, u, \ldots, u^{1-k}) +$$
$$+ \lambda^1 f(t + \tau^1, x^1, x, x^{1-2}, \ldots, x^{1-k}, u^1, u, \ldots, u^{1-k}) +$$
$$+ \mu^1 K(t + \tau^1, x^1, x, x^{1-2}, \ldots, x^{1-k}, u^1, u, \ldots, u^{1-k}).$$

$$\mathcal{H}^2 = \lambda_o f^o(t + \tau^2, x^2, x^{2-1}, x, \ldots, x^{2-k}, u^2, u^{2-1}, u, \ldots, u^{2-k}) +$$
$$+ \lambda^2 f(t + \tau^2, x^2, x^{2-1}, x, \ldots, x^{2-k}, u^2, u^{2-1}, u, \ldots, u^{2-k}) +$$
$$+ \mu^2 K(t + \tau^2, x^2, x^{2-1}, x, \ldots, x^{2-k}, u^2, u^{2-1}, u, \ldots, u^{2-k}).$$

$\vdots$

$$\mathcal{H}^k = \lambda_o f^o(t + \tau^k, x^k, x^{k-1}, \ldots, x, u^k, u^{k-1}, \ldots, u) +$$
$$+ \lambda^k f(t + \tau^k, x^k, x^{k-1}, \ldots, x, u^k, u^{k-1}, \ldots, u) +$$
$$+ \mu^k K(t + \tau^k, x^k, x^{k-1}, \ldots, x, u^k, u^{k-1}, \ldots, u).$$

$$\mathcal{H} = \sum_{r=0}^{k} {}^r.$$

$$G = \lambda_o g^o + \bar{\gamma}\bar{g} + \bar{\bar{\gamma}}\bar{\bar{g}}.$$

Now we state our conjecture:

Assume: 1) f^o, f, K and g are of class C^2, 2) $\hat{\Delta} = (\hat{u}, \hat{x}, \hat{x}_1)$ is a solution to our problem. Then there exist constants (λ_o, γ) and functions (λ, μ) with $(\lambda_o, \lambda, \mu, \gamma) \neq 0$ such that:

1.i) $\bar{\bar{\mu}} \geq 0$, $\bar{\bar{\mu}}\hat{\bar{\bar{K}}} = 0$

1.ii) $\bar{\bar{\gamma}} \geq 0$, $\bar{\bar{\gamma}}\bar{\bar{g}} = 0$

2) Euler-Lagrange-Halanay equations:

2.i) $\dot{\lambda} = -\hat{\mathcal{H}}_x$

2.ii) $\hat{\mathcal{H}}_u = 0$

3) Pontriagin maximum principle:

Setting $x, x^{r-s}, x^s = (\hat{x}, \hat{x}^{r-s}, \hat{x}^s)$ and $u^{r-s}, u^s = \hat{u}^{r-s}, \hat{u}^s$ for $s \neq r$, $s \neq 0$ in $\mathcal{H}$, $\mathcal{H}$ is maximized in u at $\hat{u}$.

4) Transversality:

$$\hat{G}_{x_1} + \lambda(t_1) = 0, \ \lambda(t_1 + \theta) = 0, \text{ for } \theta > 0.$$

2.3 <u>A control problem with a vector criterion.</u>

For the purposes of this paragraph let g^o and f^o be N-vector functions. The problem is to maximize, in the sense of Pareto, the vector of functionals:

$$J[u] = g^o(x_o, x_1) + \int_{t_o}^{t_1} f^o(t, x, u)\, dt,$$

subject to constraints 1. - 3.. Arguing exactly as we did in the finite dimensional problem, a solution, Δ, to this problem is a solution to the following N problems:

Problem $A_{\underline{i}}$: Maximize $g^{o\underline{i}} + \int_{t_o}^{t_1} f^{o\underline{i}}(t, x, u)\, dt$, subject to:

19. $g^{oi}(x_o, x_1) + \int_{t_o}^{t_1} f^{oi}(t, x, u)\, dt \geq \hat{g}^{oi}(\hat{x}_o, \hat{x}_1) + \int_{t_o}^{t_1} f^{oi}(t, \hat{x}, \hat{u})\, dt, \ i \neq \underline{i}.$

For each problem $A_{\underline{i}}$ we get a vector of multipliers $(\lambda_o^{\underline{i}}, \gamma^{\underline{i}}) \neq 0$, $(\lambda^{\underline{i}}, \mu^{\underline{i}})$ with $(\lambda_o, \lambda^{\underline{i}}, \mu^{\underline{i}}) \neq 0$, where $\lambda_o^{\underline{i}}$ is a non-negative N-vector such that conditions 1.-3. of theorem 1 of this chapter are satisfied. Summing each of these conditions over $\underline{i}$ from 1 to N and defining $\lambda_o = \sum_{\underline{i}} \lambda_o^{\underline{i}}$, $\gamma = \sum_{i} \gamma^{\underline{i}}$, $\gamma = \sum_{\underline{i}} \lambda^{\underline{i}}$, $\mu = \sum_{\underline{i}} \mu^{\underline{i}}$, we may show that: There exists a vector of non-negative constants λ_o, a vector of constants γ, and a vector function (λ, μ) such that conditions 1.-3. of theorem 1 of this chapter hold, with $\lambda_o f^o$ and $\lambda_o g^o$ interpreted as dot products.

3. A Reformulation of the Ramsey Problem in Economic Growth

3.1 Introduction.

In this section we study an economy where "resources" and produced goods are used to produce "commodities," productive capacity, and to control the growth and extraction of resources. The concept of productive capacity is used in place of capital as a factor of production and the concept of processes is used instead of production functions. The use of processes allows for joint production without assuming the absence of substitution among "factors of production" since a commodity could be produced by using any of several processes. However, the main feature of our model is the absence of the assumption of unlimited resources, which is in the background of all studies of optimal economic growth. Another feature is our attempt to solve the "original" Ramsey problem [44], which is paraphrased as follows: The community has a trade off between reaching "bliss" as soon as possible and its "welfare" during the process of reaching bliss. This trade off is represented here as a function of total utility from the beginning to the end of the "program" and the time of reaching "bliss," i.e., the time the program is completed.

We find, proposition 1, that a necessary and sufficient condition for optimality, in the above Ramsey-sense, is the existence of a price path that makes the allocations efficient. Efficiency, here, is in the following sense:

(a) A helmsman decides on the consumption paths and on the length of the planning period. The helmsman maximizes utility minus costs of consumption.

(b) With each type of capacity and with each resource, we associate an "investment manager." Investment managers maximize cumulative net returns from investment.

(c) With each production process we associate a production manager. Production managers maximize their instantaneous profits.

We also derive, proposition 3, the Ramsey investment rule, i.e., that "investment" is positive as long as the instantaneous utility is less than the bliss level. As a consequence of proposition 3 we get a comment on the "zero growth" revolution; if zero growth is optimal then we must have reached the bliss point.

In proposition 2 we prove a macroeconomic dualtiy theorem, along the lines suggested by Bruno [13]. A by-product of our duality theorem is that, in computing national income, we must subtract the depletion of resources from the value of GNP. The present way of computing GNP assumes that natural resources are free goods and would be accurate if they were.

In proposition 4, we show that if there is a feasible program then there is an optimal program. This and the other propositions are straight forward applications of theorems on the existence and characterizations of constrained extremes in the calculus of variations. The earliest existence theorem, that applies to our problem, is in McShane [35]. We shall apply theorem 1, section 1 of this chapter for necessity, theorem 2 of that section for sufficiency and of Lee and Markus [33] for existence.

2. Formulation of the Problem

2.1 Notation. Resources are classified into two basic types: the first is capital-like, in the sense that, except for decay, it is not destroyed by usage. There are L resources of type 1. The rate of growth of resource ℓ, $\ell = 1, \ldots, L$, of type 1 consists of two parts: a natural net rate of growth, a^1_ℓ, which is a given constant, and a modification rate v^1_ℓ which is controlled by the society. There will be no sign restriction on $v^1 = (v^1_1, \ldots, v^1_\ell, \ldots, v^1_L)$. We may think of type 1 resources as various types of labor. In that case v^1 will represent birth control and training efforts. The growth equations for type 1 of resources is given by:

1) $\dot{y}^1_\ell = (a^1_\ell + v^1_\ell)y^1_\ell,\ y_\ell(0) = \overset{o1}{y}_\ell > 0,$

where $y^1_\ell = y^1_\ell(t)$ is the stock of resource ℓ of type 1 at time t, t = 0 is the initial time of planning and $\overset{o1}{y}_\ell$ is a given constant representing the initial stock.

The N resources of type 2 are destroyed if used. There is a "natural" rate of growth, a^2_n, and a modification rate, v^2_n, due to recycling. We restrict v^2_n to be non-negative. Define:

0.1 $v^2 = (v^2_1, \ldots, v^2_n, \ldots, v^2_N).$

Let the rate of extracting resource n be denoted by v^3_n. The growth equations for type 2 resources are given by:

2) $\dot{y}^2_n = (a^2_n + v^2_n - v^3_n)y^2_n,\ y^2_n(0) = \overset{o2}{y}_n > 0,$

where $y^2_n = y^2_n(t)$ is the stock of resources n of type 2 at time t and $\overset{o2}{y}_n$ are initial stocks. Let $b^3_n < \infty$ be the maximum possible rate of v^3_n.

There are I produced goods in the economy and J processes of producing them. Let $g^o_{ij}(t, u_j)$ be the net outcome, in terms of good i, of operating process j at level $u_j = u_j(t)$ at time t, i = 1, ..., I; j = 1, ..., J. We shall take $g^o_{ij}(t, u_j) > 0$ to mean that good i is a net output of process j at level u_j at time t, and $g^o_{ij}(t, u_j) \leq 0$ means good i is a net output to produce j at level u_j at time t. Define the vectors $g^o_{i,j}$ and $g^o_{i,\cdot}$ as follows:

0.2 $g^o_{\cdot,j} = (g^o_{1,j}, \ldots, g^o_{I,j})$, j = 1, ..., J.

0.3 $g^o_{i,\cdot} = (g^o_{i,1}, \ldots, g^o_{i,J})$, i = 1, ..., I.

The net outcome of process j, at time t at level u_j in terms of resource of type 1 is denoted by $g^1_{\ell j}(t, u_j)$ and it is non-positive, by our convention. $g^2_{nj}(t, u_j)$ denotes the net outcome of operating process j at level u_j at time t, in terms of resource n of type 2, and it is non-positive. We define vectors $g^1_{\cdot,j}$, $g^1_{\ell,\cdot}$, $g^2_{\ell,\cdot}$, $g_{\cdot,j}$ and u as follows:

0.4 $g^1_{\cdot,j} = (g^1_{1,j}, \ldots, g^1_{L,j})$, j = 1, ..., J.

0.5 $g^1_{\ell,\cdot} = (g^1_{\ell,1}, \ldots, g^1_{\ell,j})$, ℓ = 1, ..., L.

0.6 $g^2_{,\cdot j} = (g^2_{1,j}, \ldots, g^2_{N,j})$, j = 1, ..., J.

0.7 $g^2_{n,\cdot} = (g^2_{n,1}, \ldots, g^2_{n,J})$, n = 1, ..., N.

0.8 $g_{\cdot,j} = (g^o_{\cdot,j}, g^1_{i.j}, g^2_{\cdot,j})$, j = 1, ..., J.

0.9 $u = (u_{\cdot}, \ldots, u_J)$.

We shall take u_j to be the level of operated capacity of process j. We denote the capacity of process j by x_j. The rate of change in x_j is given by:

3) $\dot{x}_j = (v^o_j - a^o_j)x_j$, $x_j(0) = \overset{o}{x}_j > 0$,

where $\overset{o}{x}_j$ is the, given, initial capacity, v^o_j is the rate of growth in capacity and a^o_j is the rate of depreciation.

The modification of resource growth rates and affecting changes in capacities of processes entails allocations of resources and produced goods. These allocations are determined by "input functions." Let $S^o_{ij}(v^o_j x_j)$, $S^1_{\cdot j}(v^o_j x_j)$, $S^2_{\cdot j}(v^o_j x_j)$ denote the inputs required to affect a growth rate v^o_j of capacity of process j, in terms of good i, resource ℓ and resource n respectively. Define:

0.10) $S^o_{i,\cdot} = (s^o_{i,1}, \ldots, S^o_{i,J})$, $i = 1, \ldots, I$.

0.11) $S^1_{\ell,\cdot} = (S^1_{\ell,1}, \ldots, S^1_{\ell,J})$, $\ell = 1, \ldots, L$.

0.12) $S^2_{n,\cdot} = (S^2_{n,1}, \ldots, S^2_{n,J})$, $n = 1, \ldots, N$.

We, further, denote the inputs to modification of type 1 of resources by $f^o_{i,\ell}(v^1_\ell, y^1_\ell)$, $f^1_{\substack{1\\ \ell,\ell}}(v^1_\ell y^1_\ell)$ and $f^2_{n,\ell}(v^1_\ell y^1_\ell)$. We define the vectors $f^o_{i,\cdot}$, $f^1_{\ell,\cdot}$ and $f^2_{n,\cdot}$ as in 0.10 - 0.12. The inputs to recycling at a rate v^2_n, of resource n, is given by $h^o_{i,n}(v^2_n y^2_n)$m $h^1_{\ell,n}(v^2_n y^2_n)$ $h^2_{\substack{1\\ n,n}}(v^2_n y^2_n)$, and the inputs to extraction, at a rate v^3_n, of y^2_n are given by $M^o_{in}(v^3_n y^2_n)$, $M^1_{\ell,n}(v^3_n y^2_n)$ and $M^2_{\substack{1\\ n,n}}(v^3_n y^2_n)$. Define $h^o_{i,\cdot}$, $h^1_{\ell,\cdot}$, $h^2_{n,\cdot}$, $M^o_{i,\cdot}$, $M^1_{\ell,\cdot}$ and $M^2_{n,\cdot}$ as in 0.10 - 0.12.

The objective of the economy is to produce goods for consumption. Let c^o_i be the consumption of good i, c^1_ℓ be consumption of resource ℓ and c^2_n be consumption of resource n. If ℓ is a type of labor then c^1_ℓ would be leisure and if n is a natural resource then c^2_n would be recreational use of that resource. Define:

0.13) $c^o = (c^o_1, \ldots, c^o_I)$, $c^1 = (c^1_1, \ldots, c^1_L)$, $c^2 = (c^2_1, \ldots, c^2_N)$.

Let the instantaneous utility function of consumption be given by $\phi(t,c)$, and define:

0.14) $x_o(t) = \int_o^t \phi(t,c)dt$.

Then we may write 0.14 as:

4) $\dot{x}_o = \phi(t,c)$, $x_o(0) = 0$.

2.2 <u>Feasibility</u>. Let $\mathcal{Z} = (y^1, y^2, x, x_o, u, v^o, v^1, v^2, v^3, c)$. We shall say that $\mathcal{Z}$ is feasible if y^1, y^2, x, x_o satisfy equations (1) - (4) with the terminal conditions: to be introduced below and if no more is used than what is available of goods and resources.

a) Availability constraints:

5.1) $E^o_i(\mathcal{Z}) = \sum_j g^o_{i,j} - \sum_j S^o_{i,j} - \sum_\ell f^o_{i,\ell} - \sum_n (h^o_{i,n} + M^o_{i,n}) - c^o_i \geq 0$, $i = 1, \ldots, I$.

5.2) $E^1_\ell(\mathcal{Z}) = y^i_\ell + \sum_j g^1_{\ell,j} - \sum_j S^1_{\ell,j} - \sum_{\ell!=1}^{L} f^1_{\ell,\ell'} - \sum_n (h^1_{\ell,n} + M^1_{\ell,n}) - c^1_\ell \geq 0$.

5.3) $E_n^2(\mathbf{z}) = v_n^3 y_n^2 + \sum_j g_{n,j}^2 - \sum_j S_{n,j}^2 - \sum_\ell f_{n,\ell}^2 - \sum_{n'=1}^{N} (h_{n,n'}^2 + M_{n,n'}^2) - c_n^2 \geq 0.$

5.4) $v_n^3 \leq b_n^3 < \infty.$

5.5) $E_j^3(\mathbf{z}) = u_j \leq x_j.$

b) Non-negativity and boundedness constraints:

6.1) $u_j \geq 0,\ 0 \leq v_j^o \leq b_n^o\ ,\ |v_\ell^1| < b_\ell^1,\ 0 \leq v_n^2 \leq b_n^2,\ v_n^3 \geq 0.$

6.2) $c_i^o \geq 0,\ c_\ell^1 \geq 0,\ c_n^n \geq 0,$

where b_j^o, b_ℓ^1, b_n^2 are given positive numbers.

c) End point conditions:

Suppose the goal of the economy is to be in a position where consumption at bliss levels is feasible. Following Ramsey [44] we interpret that to mean:

If all resources and goods and utilized processes are run at full capacity and if allocations to modify growth levels in resources and capacities are just enough to sustain a zero growth rate and if all the rest is consumed, then the maximum value, say B, of ϕ is achieved. With that interpretation, denoting the time the goal is achieved by T, we have:

(i.1) $u_j(T) = x_j(T).$

(i.2) $v_n^3(T) = b_n.$

(i.3) $v_j^o(T) = a_j^o.$

(i.4) $v^1(T) = -a_\ell^1.$

(i.5) $v_n^2(T) = -a_n^2 + v_n^3(T) = -a_n^3 + b_n^3.$

Substituting in 5.1 - 5.3, as equations, we get:

(i.6) $$c_i^o(T) = \sum_j g_{ij}^o(x_j(T)) - \sum_j S_{ij}^o(a_j^1 x_j(T)) - \sum_\ell g_{i\ell}^o(-a_\ell^1 y_\ell^1(T))$$
$$-\sum_n [h_{in}^o((b_n^3 - a_n^2)y_n^2(T)) + M_{in}^o((b_n^3 - a_n^2)y_n^2(T))].$$

(i.7) $$c_\ell^1 = y_\ell^1(T) + \sum_j g_{ij}^1(x_j(T)) - \sum_j S_{\ell j}^1\ (a_j^1 x_j(T)) - \sum_{\ell^1} f_{\ell\ell^1}^1(-a_\ell^1 y_\ell^1(T))$$
$$-\sum_n [h_{\ell n}^1((b_n^3 - a_n^2)y_n^2(T)) + M_{\ell n}^1((b_n^3 - a_n^2)y_n^2(T))].$$

(i.8) $c_n^2(T) = b_n y_n^2(T) + \sum_j g_{nj}^2(x_j(T)) - \sum_j S_{nj}^2(a_j^o x_j(T))$

$$-\sum_\ell f_{n\ell}^2(-a_\ell^1 y_\ell^1(T)) - \sum_{n'} [h_{nn'}^2((b_n^3 - a_n^2)y_n^2(T)) + M_{nn'}^2(b_n^3 - a_n^2)y_n^2(T))].$$

Denote the right hand sides of (i.6) to (i.8) by $\eta_i^o[x(T), y(T)]$, $\eta_\ell^1[x(T), y(T)]$ and $\eta_n^2[x(T), y(T)]$ respectively. The end point conditions that

(i.9) $\phi[.; \eta^o, \eta^1, \eta^2] = B$, where $\eta^o \geq 0$, $\eta^1 \geq 0$, $\eta^2 \geq 0$

and where $\phi(.,\eta)$ denotes a function that has the form of ϕ in η. Now, (i.1) - (i.8) the goal is to achieve stocks x(T) and y(T) at time T such that:

7.1) $\phi[., \eta^o(x(T) , y(T)), \eta^1(x(T), y(T)), \eta^2(x(T), y(T))] = F(\eta(T)) = B.$

7.2) $\eta^o \geq 0$, $\eta^1 \geq 0$, $\eta^2 \geq 0$.

The reason we ignore (i.1) - (i.8) is that we are not restricting c(T) but we are looking for a level of (x(T), y(T)) at time T that would make it possible to attain utility level B. Our end point conditions are, then, expressed by 7).

d. Definition of feasibility:

D.1 Definition: A program $\mathcal{Z}$ defined on [0,T] is said to be feasible if it satisfies 1) - 7).

2.3 Optimality.

Again, following Ramsey [44], there is assumed to be a trade off between achieving bliss as soon as possible and utility of consumption during the process of doing so. Let that trade off be defined by way of the function $w[x_o(T), T]$, where $x_o(t)$ is cumulative utility of consumption. We now define optimality.

D.a Definition: A program $(\hat{\mathcal{Z}}, \hat{T})$ is said to be optimal if it maximizes $w[x_o(t), T]$ among all feasible programs.

2.4 Market conditions.

Suppose $r_i^o(t)$ is the prevailing price, at time t, of produced good i, let r_ℓ^1 and r_n^2 denote returns to resources due to using their services, and let r_j^3 be the rental of capacity j. We shall call the vector $(r^o, r^1, r^2, r^3) = (r^o, r)$ a price-rental vector.

The profits that result from operating process j at level u_j at time t are given by:

8) $\Pi_j(t, r^o, r_j u_j) = r^o.g^1_{.,j} + r^2.g^2_{.,j} - r^3_j u_j.$

Let ζ be the end point of the planning horizon. We define net returns from "investment" activity, i.e., investment in capacity expansion and in modifying and extracting resources, as the cumulative rentals minus expenditures. Letting R^o_j, R^1_ℓ and R^2 denote these returns we write:

9.1) $$R^o_j(v^o_j, r^o, r, \tau) = \int_o^\zeta (r^3_j x_j - r^o.S^o_{.,j}(v^o_j x_j) - r^1.S^1_{.,j}(v^o_j x_j) - r^2.S^3_{.,j}(v^o_j x_j))dt.$$

9.2) $$R^1_\ell(v^1_\ell, r^o, r, \tau) = \int_o^\zeta (r^1_\ell y^1_\ell - r^o.f^o_{.,\ell}(v^1_\ell y^1_\ell) - r^1.f^1_{.,\ell}(v^1_\ell y^1_\ell) - r^2.f^3_{.,\ell}(v^1_\ell y^1_\ell))dt.$$

9.3) $$R^2_n(v^2_n, v^3_n, r^o, r, \tau) = \int_o^\zeta (r^2_\ell z^2_n - r^o.(h^o_{.,n}(v^2_n y^2_n) + M^o_{.,n}(v^3_n y^2_n)$$

$$-r^1.(h^1_{.,n}(v^2_n y^2_n) + M^1(v^3_n y^2_n)) - r^2.(h^2_{.,n}(v^2_n y^2_n) + M^2(v^3_n y^2_n))dt.$$

We now define an efficient program as follows:

D.3 Definition: A price-rental vector $(\hat{r}^o, \hat{r})$ and a program $(\hat{z}, \hat{T})$ are said to be an efficient program if:

1) Society maximizes utility minus total costs of consumption, i.e.,

$$w(\int_o^T \phi(t, c)dt, T) - \int_o^T (r^o.c^o + r^1.c^1 + r^2.c^2)dt$$

is maximized, at $(\hat{c}, \hat{T})$, in (c, T).

2) The profits from each production process is maximized at each point of time in $[0, T]$, i.e.,

$\Pi_j(t, \hat{r}^o, \hat{r}, u_j)$ is maximized in u_j at u_j, $\hat{u}_j \geq 0$.

3) Net returns on investment are maximized, i.e.,

3.a) $R^o_j(\hat{T}, \hat{r}^o, \hat{r}, v^o_j)$ is maximized in v^o_j, at $\hat{v}^o_j$, subject to: $\dot{x}_j = (v^o_j - a^o_j)x_j$,

$0 \leq v^o_j \leq b^o_j$.

3.b) $R^1_\ell(\hat{T}, \hat{r}^o, \hat{r}, v^1_\ell)$ is maximized in v^1_ℓ, at $\hat{v}^1_\ell$, subject to:

$\dot{y}^1_\ell = (a^2_\ell + v^1_\ell)y^1_\ell$, $|v^1_\ell| \leq b^1_\ell$.

3.c) $R^2_n(\hat{T}, \hat{r}^o, \hat{r}. v^2_n, v^3_n)$ is maximized in v^2_n, v^3_n, $\hat{v}^2_n$, $\hat{v}^3_n$, subject to:

$\dot{y}^2_n = (a^2_n + v^2_n - v^3_n)y^2_n$, $0 \leq v^2_n \leq b^2_n$, $0 \leq v^3_n \leq b^3_n$.

4) All markets are cleared, i.e.,

4.a) $E^o_i(\hat{z}) \geq 0$, $r^o_i E^o_i(\hat{z}) = 0$.

4.b) $E^1_\ell(\hat{z}) \geq 0$, $r^1_\ell E^1_\ell(\hat{z}) = 0$.

4.c) $E^2_n(\hat{z}) \geq 0$, $r^2_n E^2_n(\hat{z}) = 0$.

4.d) $E^3_j(\hat{z}) \geq 0$, $r^3_j E^3_j(\hat{z}) = 0$.

2.5 Assumptions.

In this section we list the assumptions which we use in the statements of various propositions, noting that we are not assuming all of the assumptions to hold.

A.1 The functions w, ϕ, g^o, g^1, g^2; S^o, S^1, S^2; f^o, f^1, f^2; h^o, h^1, h^2; M^o, M^1, and M^2 are continuously differentiable in their arguments.

A.2 The functions ϕ, g^o, g^1, and g^2 are concave.

A.3 Let $\xi^o_j = v^o_j x_j$, $\xi^1_\ell = v^1_\ell y_\ell$, $\xi^2_n = v^2_n y^2_n$ and let $\xi^3_n = v^3_n z_n$. We assume that the functions $S^o_{\cdot,j}$, $S^1_{\cdot,j}$, $S^2_{\cdot,j}$ are convex in ξ^o_j, the functions $f^o_{\cdot,\ell}$, $f^1_{\cdot,\ell}$, $f^2_{\cdot,\ell}$ are convex in ξ^1_ℓ, the functions $h^o_{\cdot,n}$, $h^1_{\cdot,n}$, $h^2_{\cdot,n}$ are convex in ξ^2_n and the functions $M^o_{\cdot,n}$, $M^1_{\cdot,n}$, and $M^2_{\cdot,n}$ are convex in ξ^3_n.

A.4 The functions $g^o_{\cdot,j}$, $g^1_{\cdot,j}$, $g^2_{\cdot,j}$ are homogeneous of degree one in u_j, the functions $S^o_{\cdot,j}$, $S^1_{\cdot,j}$, $S^2_{\cdot,j}$ are homogeneous of degree one in ξ^o_j, the functions $F^o_{\cdot,\ell}$, $f^1_{\cdot,\ell}$, $f^2_{\cdot,\ell}$ are homogeneous of degree one in ξ^1_ℓ, the functions $h^o_{\cdot,n}$, $h^1_{\cdot,n}$, $h^2_{\cdot,n}$ are homogeneous of degree one in ξ^2_n and the functions $M^o_{\cdot,n}$, $M^1_{\cdot,n}$, $M^2_{\cdot,n}$ are homogeneous of degree one in ξ^3_n.

A.5 The function $\phi(t, c)$ is homogeneous of degree one in c.

A.6 The function $w(x_o(T),T)$ is concave and $\frac{\partial w}{\partial x_o} > 0$.

A.7 The function ϕ and g are separable in t, i.e.,

(i) $\phi(t, C) = A_o(t)\phi(C)$

(ii) $g^o_{\cdot,j}(t, x_j) = A^o_{\cdot,j} G^o_{\cdot,j}(x_j)$

(iii) $g^2_{\cdot,j}(t, x_j) = A^1_{\cdot,j}(t) G^1_{\cdot,j}(x_j)$

(iv) $g^2_{\cdot,j}(t, x_j) = A^2_{\cdot,j}(t) G^2_{\cdot,j}(x_j)$,

where the functions A(t) are real valued and differentiable.

A.8 The functions ϕ, $|g^o|$, $|g^1|$, $|g^2|$, h^o, h^1, h^2, f^o, f^1, f^2, M^o, M^1, M^2 are monotone increasing, and, except for ϕ, the monotonicity is strict.

A.9 g^o, g^1, g^2, h^o, h^1, h^2, f^o, f^1, f^2, M^o, M^1, M^2 are zeros when and only when their respective arguments are zeros.

3. Characterization of Optimality

In this part we present necessary and sufficient conditions for optimality (section 0) and then we present some implication of these conditions (section 1) in terms of a decentralized characterization, macroeconomic duality and Ramsey-type results.

0. Necessary and sufficient conditions for optimality.

We state these conditions in the form of two theorems. The first theorem is obtained by applying theorem 1 and theorem 2, section 1 of this chapter to our problem. The second theorem is a restatement of the first theorem.

Theorem 1. If assumptions A.1, A.2, A.3, A.6, A.8, A.9 and A.10: For any $t \in [0, \hat{T}]$ at least one of $\hat{v}^o_j$, $\hat{v}^2_n$, $\hat{v}^3_n$, is positive. Then the following conditions are necessary and sufficient for the optimality of $(\hat{z}, \hat{T})$: There exists a constant λ_o and functions q_o, q^o_j, q^1_ℓ, μ^o_i, μ^1_ℓ, μ^2_n, μ^3_j, ν^o_j, ν^1_ℓ, ν^3_n, ν^4_n, γ^o_i, γ^1_ℓ, γ^2_n, γ^3_j, γ^4_n, γ^5_n, γ^6_j defined on $[0, \hat{T}]$ such that

10) $\lambda_o > 0$

11.1) $\dot{q}_o = 0$

11.2) $$\dot{q}^o_j = -(v^o_j - a^o_j)q^o_j - \mu^3_j + \mu^o \cdot \hat{v}^o_j S'^o_{\cdot,j}(\hat{v}^o_j \hat{x}_j) + \mu^1 \cdot \hat{v}^o_j S'^1_{\cdot,j}(\hat{v}^o_j \hat{x}_j) + \mu^2 \cdot \hat{v}^o_j S'^2_{\cdot,j}(\hat{v}^o_j \hat{x}_j),$$

where the components of μ^o are μ^o_i, the components μ^1 are μ^1_ℓ, the components of μ^2 are μ^2_ℓ and where the prime denotes the derivative, e.g., the components of $S'^o_{\cdot,j}(\hat{v}^o_j \hat{x}_j)$

are $$\frac{d}{d\xi^o_j} S^o_{i,j} \Bigg|_{\substack{v^o_j = \hat{v}^o_j \\ x_j = \hat{x}_j}}$$

11.3) $$\dot{q}^1_\ell = -(\hat{v}^1_\ell + a^1_\ell)q^1_\ell - \mu^1_\ell + \mu^o \cdot \hat{v}^1_\ell f'^o_{\cdot,\ell}(\hat{v}^1_\ell \hat{y}_\ell) + \hat{v}^1_\ell f'^1_{\cdot,\ell}(\hat{v}^1_\ell y^1_\ell) + \mu^2 \cdot \hat{v}_\ell f'^2(\hat{v}^1_\ell \hat{y}^1_\ell).$$

11.4) $\dot{q}_n^2 = -(a_n^2 + v_n^2 - v_n^3)q_n^2 - \mu_n^2 v_n^3 + \mu^o \cdot (\hat{v}_n^{2\prime} h^o_{\cdot,n}(\hat{v}_n^2 \hat{y}_n^2) + \hat{v}_n^{3\prime} M^o_{\cdot,n}(\hat{v}_n^3 \hat{y}_n^2))$

$+ \Gamma^1 \cdot (\hat{v}_n^2 h^1_{\cdot,n}(\hat{v}_n^2 \hat{y}_n^2) + \hat{v}_n^{3\prime} M^1_{\cdot,n}(\hat{v}_n^3 \hat{y}_n^2)) + \Gamma^2 \cdot (\hat{v}_n^{2\prime} h^2_{\cdot,n}(\hat{v}_n^2 \hat{y}_n^2) + \hat{v}_n^{3\prime} M^2_{\cdot,n}(\hat{v}_n^3 \hat{y}_n^2),$

where the components of Γ^1 are γ_ℓ^1 and the components of Γ^2 are γ_n^2.

12.1) $q_o(T) = \lambda_o \hat{w}_1$, where $\hat{w}_1 = \frac{\partial}{\partial x_o} w(x_o(T), T)\Big|_{\hat{x}_o(T),\, \hat{T}}$

12.2) $q_j^o(T) = 0^1,\ q_\ell^1(T) = 0,\ q_n^2(T) = 0.$

12.3) $q_o(T)\dot{\hat{x}}_o(T) + \sum_j q_j^o(T)\dot{\hat{x}}_j(T) + \sum_\ell q_\ell^1(T)\dot{\hat{y}}_\ell^1(T) + \sum_n q_j^2(T)\dot{\hat{y}}_n^2(T) = -\lambda_o \hat{w}_2,$

where $\hat{w}_2 = \frac{\partial}{\partial_T} w(x_o(T), T)\Big|_{\hat{T},\, \hat{x}_o(\hat{T})}$

13.1) $q_o\hat{\phi}_{oi} - \mu_i^o + \gamma_i^o = 0,\ \gamma_i^o \geq 0,\ \gamma_i^o \hat{c}_i^o = -$, where $\hat{\phi}_{oi} = \frac{\partial \phi}{\partial C_i^o}\Big|_{c = \hat{c}}$

13.2) $q_o\hat{\phi}_{1\ell} - \mu_\ell^1 + \gamma_\ell^1 = 0,\ \gamma_\ell^1 \geq 0,\ \gamma_\ell^1 \hat{c}_\ell^1 = 0$, where $\hat{\phi}_{1\ell} = \frac{\partial \phi}{\partial c^1}\Big|_{c = \hat{c}}$

13.3) $q_o\hat{\phi}_{2n} - \mu_n^2 + \gamma_n^2 = 0,\ \gamma_n^2 \geq 0,\ \gamma_n^2 \hat{c}_n^2 = 0$, where $\hat{\phi}_{2n} = \frac{\partial \phi}{\partial c_n^2}\Big|_{c = \hat{c}}$

13.4) $q_j^o \hat{x}_j - \mu^o \cdot \hat{x}_j S'^o_{\cdot,j}(\hat{v}_j^o \hat{x}_j) - \mu^1 \cdot \hat{x}_j S'^1_{\cdot,j}(\hat{v}_j^o \hat{x}_j) - \mu^2 \cdot \hat{x}_j S'^2_{\cdot,j}(\hat{v}_j^o \hat{x}_j)$

$- \nu_j^o + \gamma_j^3 = 0,\ \gamma_j^3 \hat{v}_j^o = 0,\ \nu_j^o(b_j^o - \hat{v}_j^o) = 0,\ x_j^o \geq 0,\ \gamma_j^3 \geq 0.$

13.5) $q_\ell^1 \hat{y}_\ell^1 - \mu^o \cdot \hat{y}_\ell^1 f'^o_{\cdot,\ell}(\hat{v}_\ell^1 \hat{y}_\ell^1 - \mu^1 \cdot \hat{y}_\ell^1 f'^1_{\cdot,\ell}(\hat{v}_\ell^1 \hat{y}_\ell^1) - \mu^2 \cdot \hat{y}_\ell^1 f'^2_{\cdot,\ell}(\hat{v}_\ell^1 \hat{y}_\ell^1)$

$+ \nu_\ell^1 - \nu_\ell^2 = 0,\ \nu_\ell^1(\hat{v}_\ell^1 + b_\ell^1) = 0,\ \nu_\ell^2(b_\ell^1 - \hat{v}_\ell^1) = 0,\ \nu_\ell^1 \geq 0,\ \nu_\ell^2 \geq 0.$

13.6) $q_n^2 \hat{y}_n^2 - \mu^o \cdot \hat{y}_n^2 h'^o_{\cdot,n}(\hat{v}_n^2 \hat{y}_n^2) - \mu^1 \cdot \hat{y}_n^2 h'^1_{\cdot,n}(\hat{v}_n^2 \hat{y}_n^2) - \mu^2 \cdot \hat{y}_n^2 h^2_{\cdot,n}(\hat{v}_n^2 \hat{y}_n^2)$

$- \mu_n^3 + \gamma_n^4 = 0,\ \gamma_n^4 \hat{v}_n^2 = 0,\ \nu_n^3(b_n^2 - \hat{v}_n^2) = 0,\ \nu_n^3 \geq 0,\ \gamma_n^4 \geq 0.$

13.7) $-q_n^2 \hat{y}_n^2 - \mu^o \cdot \hat{y}_n^2 M'^o_{\cdot,n}(\hat{v}_n^3 \hat{y}_n^2) - \mu^1 \cdot \hat{y}_n^2 M'^1_{\cdot,n}(\hat{v}_n^2 \hat{y}_n^2) - \mu^2 \cdot M'^2_{\cdot,n}(\hat{v}_n^3 \hat{y}_n^2)$

$- \mu \hat{y}_n^2 - \nu_n^4 + \gamma_n^5 = 0,\ \gamma_n^5 \hat{v}_n^3 = 0,\ \nu_n^4(b_n^3 - \hat{v}_n^3) = 0,\ \nu_n^4 \geq 0,\ \gamma_n^5 \geq 0.$

$$13.8)\quad \mu^{o}.\ g'^{o}_{\cdot,j}(t,\hat{u}_j) + \mu^{1}.\ g'^{1}_{\cdot,j}(t,\hat{u}_j) + \mu^{2}.\ g'^{2}_{\cdot,j}(t,\hat{u}_j) - \mu^{3}_j + \gamma^{6}_j = 0,$$

$$\gamma^{6}_j\hat{u}_j = 0,\ \gamma^{6}_j \geq 0.$$

Denoting the left hand sides of 5.1 - 5.3 evaluated at $\hat{z}$ by E^1_j, E^2_ℓ, and E^3_n,

$$14.1)\quad \mu^o_i \geq 0{:}\quad \Sigma g^o_{ij}(t,\hat{u}_j) - \Sigma_j S^o_{ij}(\hat{v}^o_j\hat{x}_j) - \Sigma_\ell f^o_i(\hat{v}^1\hat{y}^1) - \Sigma_n(h^o_{i,n}(\hat{v}^2_n\hat{y}^2_n)$$

$$+ M^o_{i,n}(\hat{v}^3_n\hat{y}^2_n)) - \hat{c}^o_i = E^o_i \geq 0,\ E^o_i\mu^o_i = 0.$$

$$14.2)\quad \mu^1_\ell \geq 0,\ y^1_\ell + \Sigma_j g^1_{ij}(t,\hat{u}_j) - \Sigma_j S^1_{ij}(\hat{v}^o_j\hat{x}_j) - \Sigma_{\ell'} f^1_{\ell,\ell'}(\hat{v}^1_{\ell'}\hat{y}^1_{\ell'})$$

$$- \Sigma_n(h^1_{\ell n}(\hat{v}^2_n\hat{y}^2_n) + M^1_{\ell n}(\hat{v}^3_n\hat{y}^2_n)) - \hat{c}^1_\ell = E^1_\ell \geq 0,\ \mu^1_\ell E^1_\ell = 0.$$

$$14.3)\quad \mu^2_n \geq 0,\ \hat{v}^3_n\hat{y}^2_n + \Sigma_j g^2_{nj}(t,\hat{u}_j) - \Sigma_\ell f^2_{n\ell}(\hat{v}^1_\ell\hat{y}^1_\ell) - \Sigma_j S^2_{nj}(\hat{v}^o_j\hat{x}_j) - \Sigma_{n'}(h^2_{n,n'}(\hat{v}^2_{n'}\hat{y}^2_{n'})$$

$$+ M^2_{n,n'}(\hat{v}^3_{n'}\hat{y}^2_{n'})) - \hat{c}^3_n = E^3_n \geq 0,\ \mu^3_n E^3_n = 0.$$

$$14.4)\quad \mu^3_j \geq 0,\ \hat{x}_j - \hat{u}_j \geq 0,\ \mu^3_j(\hat{x}_j - \hat{u}_j) = 0.$$

Proof:

Theorem 1 is proved by applying theorem 1, section 1 of this chapter to obtain the necessity part and theorem 2 of section 1 to obtain sufficiency.

a) Necessity: By A.1 and the fact that the Jacobian of the constraints (5) with respect to u, v, c is non-singular, theorem 1 of section 1 of this chapter applies. Conditions (11)-(14) of our theorem follow directly from conditions (3.1), (3.2) and (3.3) of Hestenes. However, the following two points need clarification:

a.1) $\lambda_o > 0$, (a.2) $q^o_j(T) = 0$, $q^1_\ell(T) = 0$, $q^2_n(T) = 0$, since it is not obvious how they follow from Hestenes theorems. We now clarify these points.

(a.1) $\lambda_o > 0$.

Hestenes states that $\lambda_o \geq 0$ and not all multipliers vanish simultaneously. To show that $\lambda_o > 0$ assume, by way of contradiction, that $\lambda_o = 0$. We shall show that this implies that all other multipliers are zeros at a given point, namely $t = \hat{T}$. This contradicts the assertion of theorem 1, section 1 of this chapter. By 12.1) $\lambda_o = 0$ implies $q_o(\hat{T}) = 0$. But, by 12.2) $q^o_j(\hat{T}) = 0$, $q_\ell(\hat{T}) = 0$ and $q^2_n(\hat{T}) = 0$. It remains to show that $\mu^\alpha(\hat{T})$, $\alpha = 0, 1, 2, 3$, $\nu^\beta(\hat{T}) = 0$, $\beta = 0, 1, 2, 3, 4$ and $\gamma^\delta(\hat{T}) = 0$, $S = 0, 1,$

2, ..., 6. By A.10, one of the $\gamma_j^3(\hat{T})$'s, the $\gamma_n^4(\hat{T})$'s, or the $\gamma_n^5(\hat{T})$'s is zero. In the case where $\gamma_{\bar{j}}^3(\hat{T}) = 0$, $j \in \{1, \ldots, J\}$, we have, by 13.4): a non-negative linear combination of μ^o, μ^1, μ^o, $\nu_{\bar{j}}^o$ is zero. Hence each term is zero. Thus $\nu_{\bar{j}}^o(\hat{T}) = 0$, since its coefficient is one. But $\hat{x}_{\bar{j}}(\hat{T}) > 0$, from equation (3), and $S'^{\alpha}_{\cdot,\bar{j}}(\hat{v}_{\bar{j}}^o\hat{x}_{\bar{j}}) > 0$, by A.1 and A.8 $= 0, 1, 2$. Thus $\mu^o = o$, $\mu^1 = 0$, $\mu^2 = 0$. This implies, by 13.1), that $\gamma^{\beta}(T) = 0$, $\beta = 0, 1, 2$. Substituting for μ^{α}, $\alpha = 0, 1, 2$, in 13.4) for $j \neq \bar{j}$, we have $\nu_j^o = \gamma_j^3 = 0$ (for $\nu_j^o > 0$ implies $\hat{v}_j^o = b_j^o > 0$ which implies $\gamma_j^3 = 0$). Substituting in 13.5), for μ^{α}, $\alpha = 0, 1, 2$, we have $v_{\ell}^1 = v_{\ell}^2$. But again, this is possible only if $v_{\ell}^1(\hat{T}) = v_{\ell}^2(\hat{T}) = 0$. Substituting in 13.6), we get $\nu_n^3 = \gamma_n^4$ which implies $\gamma_n^3(\hat{T}) = \gamma_n^4(\hat{T}) = 0$. Also, by 13.7), we have $\nu_n^4(\hat{T}) = \gamma_n^5(\hat{T}) = 0$. From 13.8) we get $\mu_j^3(\hat{T}) = \gamma_j^6(\hat{T})$. But if $\hat{u}_j(\hat{T}) = 0$ then $\hat{u}_j(\hat{T}) < \hat{x}_j(\hat{T})(\hat{x}_j(\hat{T}) > 0)$ and $\mu_j^3(\hat{T}) = \gamma_j^6(\hat{T}) = 0$. And if $\hat{u}_j(\hat{T}) > 0$ then $\gamma_j^6(\hat{T}) = 0$ and $\mu_j^3(\hat{T}) = 0$.

By arguing as above we can establish that μ^{α}, ν^{β} and γ^{δ} are zeros at $\hat{T}$ for the case where $\hat{v}_{\ell}^2(\hat{T}) > 0$ and for the case where $\hat{v}_{\ell}^3(\hat{T}) > 0$. Thus a.1 is established.

(a.2) $q_j^o(T) = 0$, $q_{\ell}^1(T) = 0$, $q_n^2(T) = 0$.

From Hestenes' theorem, it follows that $q_j^o(T) = \left.\frac{\partial}{\partial x_j(T)} G\right|_{\hat{z}}$,

$q^1(T) = \left.\frac{\partial}{\partial y_{\ell}^1(T)} G\right|_{\hat{z}}$, $q_n^2(T) = \left.\frac{\partial}{\partial y_n^2(T)} G\right|_{\hat{z}}$, where G in our case is $-\lambda_o w + \lambda_1 F[\eta] + K_1\eta_1 + K_2\eta_2 + K_3\eta_3$, where λ_1, K_1, K_3 are constants. But, by feasibility, we know that ϕ is maximized, in $\xi(T)$, $y^1(T)$, $y^2(T)$, $\hat{x}(T)$, $\hat{y}^1(T)$, $\hat{y}^2(T)$, subject to $\eta \geq 0$. Thus $\left.\frac{G}{\partial x(T)}\right|_{\hat{x},\hat{y}} = \left.(\lambda_1 \frac{\partial F}{\partial x} + K_1 \frac{\partial \eta_1}{\partial x} + K_2 \frac{\partial \eta_2}{\partial x} + K_3 \frac{\partial \eta_3}{\partial x})\right|_{\hat{x},\hat{y}} = 0$. $\left.\frac{\partial G}{\partial y(T)}\right|_{\hat{x},\hat{y}}$ $= \left.(\lambda_1 \frac{\partial F}{\partial y} + K_1 \frac{\partial \eta_1}{\partial y} + K_2 \frac{\partial \eta_2}{\partial y} + K_3 \frac{\partial \eta_3}{\partial y})\right|_{\hat{x},\hat{y}} = 0$. This establishes a.2. The other conditions follow from differentiating the Hamiltonian, with respect to x_o, x and y to get 11) and with respect to c, v, u to get 13. 12.1) and 12.3) follow from differentiating G with respect to $x_o(T)$ and from equating the Hamiltonian at $\hat{T}$ to derivative of G with respect to T.

b) <u>Sufficiency</u>: Theorem 2 of section 1 of this chapter applies since the concavity and smoothness assumptions are satisfied by A.1, A.2 and A.3 and since q_o,

q^o_j, q^1_ℓ and q^2_n are non-negative, as we show now. Multiply (13.1) by $\hat{c}^o_i$. Then $q_o \hat{c}^o_i \hat{\phi}^1_{oi} = \mu^o_i \geq 0$. Since $\hat{c}^o_i \geq 0$ and $\hat{\phi}^i_{oi} \geq 0$, by A.1 and A.8, $q_o \geq 0$. Multiplying (13.4) by $\hat{v}^o_i$ we get $q^o_j \hat{v}^o_j \hat{x}_j \geq 0$. But $\hat{v}_j \geq 0$ and, by (14.4), $\hat{x}_j \geq 0$. Thus $q^o_j \geq 0$. Multiplying (13.5) by ν^2_ℓ, and noting that $\nu^1_\ell \nu^2_\ell = 0$, we have $q^1_\ell y^1_\ell \nu^2_\ell \geq 0$. Thus $q^1_\ell > 0$. Multiplying (13.6) by $\hat{v}^2_n$ we get $q^2_n \hat{y}^2_n \hat{v}^2_n \geq 0$, thus $q^2_n \geq 0$.

Theorem 2. If assumptions A.1 - A.3 and A.6 - A.10 hold then the following conditions are necessary and sufficient for the optimality of $(\hat{Z}, \hat{T})$:

There exist functions $\mu^1(t)$, $\mu^1(T)$, $\mu^2(T)$ and $\mu^3(T)$ defined on $[0, \hat{T}]$ such that, for all $t \in [0, \hat{T}]$,

(15) $w[\int^T \phi(t, c)\, dt, T] - \int_o^T (\mu^o, c^o - \mu^1 . c^1 - \mu^2 . c^2) dt$ is maximized in (c, T) at $(\hat{c}, \hat{T}^o)$ subject to $c \geq 0$.

(16) $R^o_j[v^o_j, \mu^o, \mu^1, \mu^2, \mu^3, \hat{T}]$ is maximized in v^o_j, at $\hat{v}^o_j$, subject to:
$\dot{x}_j = (v^o_j - a^o_j) x_j$, $x_j(0) = x^o_j$, $0 \leq v^o_j \leq b^o_j$, where R^o_j is defined by (9.1).

(17) $R^1[v^1_\ell, \mu^o, \mu^1, \mu^2, \hat{T}]$ is maximized in v^1_j at $\hat{v}^1_j$, subject to:
$\dot{y}^1_\ell = (v^1_\ell + a^1_\ell) y^1_\ell$, $y^1_\ell(0) = \overset{o1}{y}_\ell$, $|v^1_\ell| \leq b^1_\ell$.

(18) $R^2_n[v^2_n, v^3_n, \mu^o, \mu^1, \mu^2, \hat{T}]$ is maximized in v^2_n, v^3_n, at $\hat{v}^2_n$, $\hat{v}^3_n$, subject to:
$\dot{y}^2_n = (a^2_n + v^2_n - v^3_n) y^2_n$, $y^2_n(0) = \overset{o2}{y}_n$, $0 \leq v^2_n \leq b^2_n$, $0 \leq v^3_n \leq b^3_n$.

(19) $\Pi_j(t, \mu^o, \mu^1, \mu^2, \mu^3, u_j)$ is maximized in u_j, at u_j, subject to $u_j \geq 0$.

(20) Relations (14) of theorem 1 hold.

Proof of theorem 2:

We establish theorem 2 by showing that conditions (15) - (19) are equivalent to conditions (10) - (13). Then theorem 2 follows from theorem 1.

a) (10) - (14) imply (15) - (19).

Take λ_o, in (10), equal to one, since $x_o > 0$. Solving (11.1) with (12.1) as terminal conditions we have

(21) $q_o(t) = w_1(\hat{x}_o, \hat{T})$.

Substituting in (13.1) - (13.3) we have the first order necessary conditions for maximizing $w(x_o(T), T) - \int_o^T \mu . c dt$ subject to $c \geq 0$, $\dot{x}_o = \phi(t, c)$, $x_o(0) = 0$. By concavity of ϕ and w, these are also sufficient and we get (15), by applying Mangasarian's theorem 1 [6]. That (13.8) imply (19) follows from sufficiency theorems for ordinary maximization.

To obtain (16) we observe:

(a.3) Remark: The following are necessary and sufficient for $\hat{v}_j^o$, and hence $\hat{x}_j^o$, to be a solution to (16):

There exists a constant $\bar{\lambda}_o > 0$ and functions $p_j(t)$, $\theta_j^1(t)$, $\theta_j^2(t)$ defined on $[0, \hat{T}]$ such that

(a.3.i) $\dot{p}_j(t) = \bar{\lambda}_o(\mu^o.\hat{v}_j^o S^o_{\cdot,j}(\hat{v}_j^o\hat{x}_j) = \mu^1.\hat{v}_j^o S'^1_{\cdot,j}(\hat{v}_j^o\hat{x}_j) + \mu^2.\hat{v}_j^o S'^2_{\cdot,j}(\hat{v}_j^o\hat{x}_j)) -$

$$p_j(\hat{v}_j^o - a_j^1) - \mu_j^3.$$

(a.3.ii) $p_j(\hat{T}) = 0.$

(a.3.iii) $\bar{\lambda}_o(-\mu^o.\hat{x}_j S'^1_{\cdot,j}(\hat{v}_j^o\hat{x}_j) - \mu^1.\hat{x}_j S'^1_{\cdot,j}(\hat{v}_j^o\hat{x}_j) - \mu^2.\hat{x}_j S'^2_{\cdot,j}(\hat{v}_j^o\hat{x}_j)) + p_j\hat{x}_j +$

$$\theta_j^1 - \theta_j^2 = 0; \ \theta_j^1\hat{v}_j^o = 0, \ \theta_j^2(b_j^o - \hat{v}_j^o) = 0.$$

The necessity part of the remark follows from Hestenes' theorem (3.1) [3], if we note $\bar{\lambda}_o \neq 0$ (if $\bar{\ }_o = 0$ then, since $p_j(\hat{T}) = 0$, we have, by (a.3.iii) $\theta_j^1(T) = \theta_j^2(T)$ which is true only when $\theta_j^1(T) = \theta_j^2(T) = 0$). The sufficiency follows, by A.2 and A.3, from theorem 2 of section 1 of this chapter.

Now (16) follows from (11.2), (12.2) and (13.4) by taking $q_j = p_j$ and applying the sufficiency part of the remark a.1. In the same manner, (17) follows from (11.3), (12.2) and (13.5) and (18) follows from (11.4), (12.2), (13.6) and (13.7).

b) Conditions (15) - (19) imply conditions (10) - (14).

Take $\lambda_o = 1$ then (10) is satisfied. Take $q_j^o = p_j$ in remark (a.3), then by the necessity part of the remark (16) implies (11.2), the first part of (12.2) and (13.4). In a similar way we may show that (11.3), (11.4), the rest of (12.2), (13.5), (13.6) and (13.7) follow from (17) and (18). (11.1) follows by taking $q_o(t) = \hat{w}_1$. (12.3) and (13.1) - (13.3) follow from (15) by first order necessary conditions of theorem 1 section 1. Conditions (13.8) follow from (19), by first order necessary conditions for ordinary, finite, extrema. This completes the proof of theorem 2.

(a.4) Remark: Solving equations (11) of theorem 1, with (12) in mind, we have:

(22.0) $q_o(t) = w_1$

(22.1) $q_j^o(t) = \dfrac{1}{\hat{x}(t)} \int_t^T (\mu^3(s) - \mu^o.\hat{S}'^o_{\cdot,j}\hat{v}_j^o - \mu^1\hat{S}'^1_{\cdot,j}\hat{v}_j^o - \mu^2\hat{S}'^2_{\cdot,j}\hat{v}_j^o)\,\hat{x}_j(s)ds,$

where the ^ signifies the fact that the function is evaluated at Z(s),

(22.2) $q^1(t) = \frac{1}{y^1_\ell(t)} \int_t^T (\mu^1_\ell - \mu^o.\hat{f}'^o_{.,\ell}\hat{v}^1_\ell - \mu^1.\hat{f}'^1_{.,\ell}\hat{v}^1_\ell - \mu^2\hat{f}'^2_{.,\ell}\hat{v}^1_\ell)\, \hat{y}^1_\ell(s)ds.$

(22.3) $q^2_n(t) = \frac{1}{\hat{y}^2_n(t)} \int_t^T [\mu^2_n\hat{v}^3_n - \mu^o.(\hat{v}^2_n\hat{h}'^o_{.,n} + \hat{v}^3_n\hat{M}'^o_{.,n}) - \mu^1.(\hat{v}^2_n\hat{h}'^1_{.,n} + \hat{v}^3_n\hat{M}'^1_{.,n})$

$- \mu^2.[\hat{v}^2_n\hat{h}'^2_{.,n} + \hat{v}^3_n\hat{M}'^2_{.,n})]\, y^2_n(s)ds.$

(a.5) Remark: A further necessary condition for optimality of $(\hat{Z}, T)$ is:

(23) $\hat{w}_1\phi(t, \hat{c}) + q^o.\dot{\hat{x}} + q^1.\dot{\hat{y}}^1 + q^2.\dot{\hat{y}}^2 = -\hat{w}_2 - \int_t^T [w_1 \frac{\partial}{\partial s}\phi(s, c) + \sum_{i,j} \mu^o_i \frac{\partial}{\partial s} g^o_{ij}(s, \hat{u}_j)$

$+ \sum_{\ell,j} \mu^1 \frac{\partial}{\partial s} g^1_{\ell,j}(s, \hat{u}_j + \sum_{i,j} \mu^2_n \frac{\partial}{\partial s} g^2_{n,j}(s, \hat{u}_j)]ds.$

Remark (a.4) follows by writing down the form of solution for the linear, in q, equations (11) and taking into consideration (12) and (1) - (3). Remark (a.5) follows from theorem 1 of section 1 of this chapter, by using (12.3) to evaluate the constant of integration.

(a.6) Remark: $\mu^\alpha(T) = 0$, $\alpha = 0,1,2$.

This is obvious, since, in the course of proving that $\lambda_o > 0$, in the proof of theorem 1 we did not use the contradiction assumption that $-\lambda_o = 0-$ in showing that $\mu^\alpha(T) = 0$.

1. Economic theoretical propositions

Proposition 1. Decentralized characterization. If A.1 - A.3, A.6, A.8 - A.10 hold, then a program $[\hat{Z}, \hat{T}]$ is optimal if and only if there exist a prices rental vector (r^o, r^1, r^2, r^3), defined on $[0, \hat{T}]$, such that $[\hat{Z}, \hat{T}]$ is an efficient allocation, in the sense of definition D.3.

Proof: The proposition follows from theorem 2 by taking $r^o = \mu^o$, $r^1 = \mu^1$, $r^2 = \mu^2$ and $r^3 = \mu^3$. (2), (3) and (4) in D.3 are, then, equivalent to (19), (16) - (18) and (20) of theorem 2, respectively. (1) in the definition D.3 is equivalent to (15) in theorem 2.

Proposition 2. Aggregate duality and accounting. If, in addition to the assumptions of proposition 1, A.4 holds, $(\hat{Z}, \hat{T})$ is optimal and A.11 $\hat{v}^o_j < b^o_j$, $|v^1_\ell|$

$< b^1_\ell, \hat{v}^2_n < b^2_n, \hat{v}^3_n < b^2_n.$

Then there exists a price rental path (r^o, r^1, r^2, r^3) such that, in addition to the efficiency of $(\hat{Z}, \hat{T})$ we have:

The weighted sum of gross rates of growth in capacities + the weighted sum of rates of reduced growth in resources - the value of depletion of resource 2 + the value of consumption = The returns on capacities + returns to resources = Gross national income, where the weights of rates of growth are the imputed values of the stocks, i.e.,

$$\sum_j \bar{q}^o_j\left(\frac{\dot{x}_j}{\hat{x}_j}\right) + \sum_\ell \bar{q}^1_\ell\left(\frac{\dot{y}^1_\ell}{\hat{y}^1_\ell}\right) + \sum_n \bar{q}^2_n\left(\frac{\dot{y}^2_n}{\hat{y}^2_n}\right) + r^o.\hat{c}^o + r^1.\hat{c}^1 + r^2.\hat{c}^2 - \sum_n r^2_n\hat{v}^3_n\hat{y}^2_n = r^1.\hat{y}^1 + r^3.\hat{x}$$

$+ \sum_n r^2_n\hat{v}^3_n\hat{y}^2_n$, where $\dot{x}_j = \hat{v}^o_j\hat{x}_j = \dot{x}_j + a^o_j x_j$, $\dot{y}^1 = \hat{v}^1\hat{y}^1$, $\dot{y}^2_n = (\hat{v}^2_n - \hat{v}^3_n)\,\hat{y}^2_n$, $\bar{q}^o_j =$

$\int_t^{\hat{T}}(r^3_j\hat{x}_j - r^o.\hat{S}^o_{.,j} - r^1.\hat{S}^1_{.,j} - r^2.\hat{S}^2_{.,j})ds$, $\bar{q}^1_\ell = \int_t^{\hat{T}}(r^1_\ell y^1_\ell - r^o.\hat{f}^o_{.,\ell} - r^1.\hat{f}^1_{.,\ell} -$

$r^2.\hat{f}^2_{.,\ell})ds$ and $\bar{q}^2_n = \int_t^T(r^2_n\hat{v}^3_n\hat{y}^2_n - r^o.(\hat{h}^o_{.,n} + \hat{M}^o_{.,n}) + r^1.(\hat{h}^1_{.,n} + \hat{M}^1_{.,n}) + r^2.(\hat{h}^2_{.,n} +$

$\hat{M}^2_{.,n})ds.$

Proof: First we note that the $\bar{q}$'s are imputed prices since they are expressed as the cumulative net returns on investment. The proposition follows from theorem 1 above. Taking, as in proposition 1, $r = \mu$ and multiplying both sides of equation (13.4) - (13.8) by $\hat{v}^o_j, \hat{v}^1_j, \hat{v}^2_j, \hat{v}^3_j$, respectively, we get, in view of the homogeniety and the vanishing $\nu^o_j, \nu^1_\ell, \nu^2_\ell, \nu^3_n, \nu^4_n$ - by assumption of this proposition -

(24) $q^o_j\hat{x}_j\hat{v}^o_j = r^o.\hat{S}^o_{.,j} + r^1.\hat{S}^1_{.,j} + r^2\hat{S}^2_{.,j}$

(25) $q^1_\ell\hat{y}^1_\ell\hat{v}^1_\ell = r^o.\hat{f}^o_{.,\ell} + r^1.\hat{f}^1_{.,\ell} + r^2.\hat{f}^2_{.,\ell}$

(26.1) $q^2_n\hat{y}^2_n\hat{v}^2_n = r^o.\hat{h}^o_{.,n} + r^1.\hat{h}^1_{.,n} + r^2.\hat{h}^2_{.,n}$

(26.2) $-q^2_n\hat{y}^2_n\hat{v}^3_n = r^o.\hat{M}^o_{.,n} + r^1.\hat{M}^1_{.,n} + r^2.\hat{M}^2_{.,n} + \sum_n r^2_n\hat{v}^3_n\hat{y}^2_n$

(27) $\mu^3_j\hat{u}_j = r^o.\hat{g}^o_{.,j} + r^1.\hat{g}^1_{.,j} + r^2.\hat{g}^2_{.,j}$

Summing each set of equations (24) - (27), adding the sums of (24) - (26) and using (14) we get:

(28.1) $\sum_j q^o_j \hat{x}_j \hat{v}^o_j + \sum_\ell q^1_\ell \hat{y}^1_\ell \hat{v}^1_\ell + \sum_n q^2_n(\hat{v}^2_n - \hat{v}^3_n)\hat{y}^2_n =$

$\sum_j (r^o.\hat{g}^o_{.,j}) - r^o.\hat{c}^o + \sum_j (r^1.\hat{g}^1_{.,j}) - r^1.\hat{c}^1 + \sum_j (r^2.\hat{g}^2_{.,j}) - r^2.\hat{c}^2 + 2\sum_n r^2_n \hat{v}^3_n \hat{y}^2_n$

(28.2) $\sum_j r^3_j \hat{u}_j = \sum_j r^3_j \hat{x}_j = \sum_j (r^o.\hat{g}^o_{.,j} + r^1.\hat{g}^1_{.,j} + r^2.\hat{g}^2_{.,j})$

From (28.1) and (28.2) we have, using the notation of this proposition for $\dot{\bar{x}}_j$, $\dot{\bar{y}}^1_\ell$, $\dot{\bar{y}}^2_n$, we have

(29) $\sum_j q_j \dot{\bar{x}}_j + \sum_\ell q_\ell \dot{\bar{y}}^1_\ell + \sum_n q_n \dot{\bar{y}}_n + r^o.\hat{c}^o + r^1.\hat{c}^1 + r^2.\hat{c}^2 - \sum_n r^3_n \hat{v}^3_n \hat{y}^2_n =$

$r^3.\hat{x} + r^1.\hat{y}^1 + \sum_n r^2_n \hat{v}^3_n \hat{y}^2_n.$

The assertion, now, follows from (29), (22.1), (22.2) and (22.3) by noting the homogeniety of S, f, h, and M.

Proposition 3. Ramsey-type results. If in addition to the assumption of proposition 2, A.7 holds with A_o, $A^\alpha_{.,j}$, $\alpha = 0, 1, 2$, as constants $\bar{A}_o$, $\bar{A}^\alpha_{.,i}$, then

(1) $\hat{c}(\hat{T})$ is indifferent to $c_o(T)$ where $c_o(T)$ is the consumption level corresponding to zero optimal rates of growth at $\hat{T}$, i.e., to $\dot{x}(\hat{T}) = 0$, $\dot{y}(\hat{T}) = 0$ (see (i.1) - (i.9) in 2.2C above), i.e., $\hat{c}(\hat{T})$ gives ϕ the same value, namely B, as the zero optimal terminal growth. In other words, bliss is achieved.

(2) Investment equation: For any $t \in [0,T]$, the weighted sum of net rates of growth has the same sign as: bliss level utility minus utility of optimal consumption where the weights are imputed prices of capacity and resources. I.e.,

$$(30) \quad \sum_j \bar{q}^o_j \frac{\dot{\hat{x}}_j}{\hat{x}_j} + \sum_\ell q \frac{\dot{\hat{y}}^1_\ell}{\hat{y}^1_\ell} + \sum_n \bar{q}^2_n \frac{\dot{\hat{y}}^2_n}{\hat{y}^2_n} = A_o w_1 (B - \phi(\hat{c}(t))), \text{ for } t \in [0,\hat{T}],$$

where $\bar{q}$ is as in proposition 2.

Proof: (1) follows from remark (a.6) and conditions (13.1) - (13.3) of theorem 1, by concavity of ϕ in c. (2) follows from remark (a.5), by observing that the integrals n (23) are zeros and using (12.3) of theorem 1. We then have:

$$(31) \quad \hat{w}_1\hat{\phi} + \sum_\ell \bar{q}^o_j(\frac{\dot{\hat{x}}_j}{\hat{x}_j}) + \sum_\ell \bar{q}^1 \frac{\dot{\hat{y}}^1_\ell}{\hat{y}^1_\ell} + \sum_n \bar{q}^2_n \frac{\dot{\hat{y}}^2_n}{\hat{y}^2_n} = -\hat{w}_2 = \hat{w}_1\phi[\hat{c}(\hat{T})]$$

But the right hand side of (31) equals B, by (1) of this proposition. This establishes (2).

Notes.

1) Assertion (2) of proposition (2) corresponds to Ramsey's result (5) in [7].

2) If, in the hypothesis of proposition 2, A_o and A^{α}_{ij} are not constants then, letting I_2 denote the weighted sum of growth rates, we have:

$$(32) \quad I_2 = w_1[B - A\hat{\phi}] - \int_t^T (\dot{A}_o \hat{w}_1 \hat{\phi} + \sum_{i,j} \dot{A}^o_{i} \mu^o_{i} \hat{G}_{ij} + \sum_{\ell,\ell} \dot{A}^1_{\ell j}\, {}^1G_{\ell j} + \sum_{n,j} \dot{A}^2_{nj} \mu^2_n \hat{G}_{nj}) ds).$$

Comparing (32) with (30) we get the effects of technological change on the growth rate that has to prevail in order to achieve bliss. Unfortunately, the two equations are not directly comparable, for the paths of consumption may not be the same. Let $\hat{z}_1$ denote the optimal path in the absence of technological change, let $\hat{z}_2$ denote the optimal path in the presence of technological change, and let I_1 denote the left hand side of (30). Then

$$(33) \quad I_1 - I_2 = \hat{w}_1[\hat{x}_{o2}, \hat{T}_2]A_o\phi[\hat{c}_2] - w_1[\hat{x}_{o1}, \hat{T}_1]\bar{A}_o\phi[\hat{c}_1] - \int_t^{\hat{T}_2} A_o \hat{w}_1[\hat{x}_{o2}, \hat{T}]\, \phi[\hat{c}_2] ds$$
$$- \int_t^{\hat{T}_2} \theta ds,$$

where θ denotes the last three terms of the integrand in (32). The first three terms represent the difference in utility between $\hat{c}_2$ and $\hat{c}_1$. If this difference is zero or negative, and if technological change is beneficial, i.e., if $\theta \geq 0$, then $I_1 - I_2 \leq 0$, i.e., the weighted sum of the optimal rates of growth are higher in the presence of technological change.

3) Under the assumptions of proposition 2, if the optimal path is the zero growth path from some point t_o on, then $\phi(t, \hat{c}) = B$, $t \in [t_o, \hat{T}]$. This follows from (30) above.

4. Existence of Optimal Programs

In this section we establish the existence of an optimal path for the finite horizon case, under our assumptions, for the Problem formulated in 2. The proof of existence consists merely of verifying the conditions of Lee-Markus' theorem 4 [33] for optimal control. We define a horizon of length τ with τ as an upper bound on T and $\tau < \infty$.

Proposition 4. If A.1, A.2, A.3 and if

(i) c, u, v are measurable on [0, τ],

(ii) There exists a feasible program,

(iii) ϕ is monotone increasing in c,

there exists an optimal program.

Proof: We shall show that assumption 1-5 and (a), (b), (c) of theorem 4, chapter 4 of Lee and Markus [33] are satisfied. Assumption 1, initial and terminal sets are compact and continuous, is satisfied, since the initial set, set of x_o, x, y that satisfy the initial conditions is fixed and since the terminal set is continuous. The continuity of the terminal set follows from the continuity of the functions defining it, i.e., by A.1 and (7) in the definition of feasibility. The compactness of the terminal set will be established below (assertion *).

2., the control set, i.e., the set of c, v, u that satisfy (5) and (6) is:

(i) Non-empty

(ii) compact for a given x_o, x, y, t

(iii) continuous.

(i) is satisfied by assumption (ii) of the proposition, i.e., by the existence of a feasible program. (ii) by (6) c, u, v are bounded below and v is bounded above by (5) c and u are bounded above. That closedness of the control set follows from the continuity of the functions in (5) and from the closedness of the set of c, u, v that satisfy (6).

(iii) follows from the continuity of the functions in (5) and (6).

3., continuity of state constraints, is satisfied due to the absence of such constraints in our problem, i.e., state constraints are identity functions, hence are continuous.

4., admissible controls are measurable, is satisfied by assumpt on (i) of the proposition.

5., continuity of the "cost functional," is satisfied since "f^o" = 0 in our problem and since w is continuous by A.1.

(a), the family of admissible controls is not empty, is satisfied by (i) and (ii) of our proposition.

That (b), there exists a uniform bound on the state variables, is satisfied is now demonstrated.

By (1), (2), (3), (4) and (6)

(b.1) $\dot{y}^1_\ell \leq (a^1_\ell + b^1_\ell)y^1_\ell$

(b.2) $\dot{y}^2_n \leq (a^2_n + b^2_n - b^3_n)y^2_n$

(b.3) $\dot{x}_j \leq (b^o_j - a^o_j)x_j$

(b.4) $\dot{x}_o = \phi(t, c) \leq \int_o^t \phi(S, c)ds = x_o(t)$

Thus there exist a positive constant K such that $x_o\dot{x}_o + \sum_j x_j\dot{x}_j + \sum_\ell y^1_\ell \dot{y}^1_\ell + \sum_n y^2_n \dot{y}^2_n \leq K$ $(||(x_o, x, y^1, y^2)|| + 1)$ for every $t \in [0, \tau]$ and for all feasible c, u, v. Following Cezari [2], this establishes the uniform boundedness of the state variables. We also have:

* The terminal set is compact. The boundedness of the terminal set follows from the uniform boundedness of the "state variables" $x_o(T)$, $x(T)$, $y(T)$. The closedness of the terminal set follows from the continuity of the functions in (7), in the definition of feasibility. We now show that c, the set $\{\dot{x}_o, \dot{x}, \dot{y}$ for feasible c, u, v$\}$ = $V(t, x_o, x, y)$ is convex for every given t, x_o, x, y, is satisfied. Let $\bar{p} = (\dot{\bar{x}}_o, \dot{\bar{x}}, \dot{\bar{y}})$ and $\bar{\bar{p}} = (\dot{\bar{\bar{x}}}_o, \dot{\bar{\bar{x}}}, \dot{\bar{\bar{y}}})$ be in V. We shall show that $\tilde{p} = \alpha\bar{p} + (1-\alpha)\bar{\bar{p}} \in V$ for $0 \leq \alpha \leq 1$. $\bar{p} \in V$ if there exists a point $\bar{c}$, $\bar{u}$, $\bar{v}$ satisfying (5) and (6) such that $\bar{p}$ is given by (1) - (4) at that point. $\bar{\bar{p}} \in V$ if there exists a point $\bar{\bar{c}}$, $\bar{\bar{u}}$, $\bar{\bar{v}}$ satisfying (5) and (6) such that $\bar{\bar{p}}$ is given by (1) - (4) at that point. To show that $\tilde{p} \in V$ we must show that there exists a point $\tilde{C}$, $\tilde{u}$, $\tilde{v}$ satisfying (5) and (6) such that $\tilde{p}$ is given by equations (1) - (4) at that point. We claim that $\tilde{c}$, $\tilde{u}$, $\tilde{v}$ = $\alpha(c^o, \bar{u}, \bar{v}) + (1 - \alpha)(c^o, \bar{\bar{u}}, \bar{\bar{v}})$, where $0 \leq c^o \leq \alpha\bar{c} + (1 - \alpha)\bar{\bar{c}}$. In fact $(\tilde{c}, \tilde{u}, \tilde{v})$ satisfies (5) and (6). It follows from the concavity of the left hand side of (5) and (6) that $\alpha(\bar{c}, \bar{u}, \bar{v}) + (1 - \alpha)(\bar{\bar{c}}, \bar{\bar{u}}, \bar{\bar{v}})$ is in V which implies that $(\tilde{c}, \tilde{u}, \tilde{v}) \in V$ because of the way c enters in (5.1) - (5.3) and (6). It remains to show that $\tilde{p}$ is given by the equations (1) - (4) at $(\tilde{c}, \tilde{u}, \tilde{v})$. By linearity of equations (1) - (3) in $\dot{v}$, $\dot{x}$, $\dot{\tilde{y}}^1$, y^2 is given by (1) - (3) at $(\tilde{c}, \tilde{u}, \tilde{v})$. The last question is: does there exist a point $c^o \leq \alpha\bar{c} + (1 - \alpha)\bar{\bar{c}} = \tilde{\tilde{c}}$, say, such that $\alpha\dot{\bar{x}}_o + (1 - \alpha)\dot{\bar{\bar{x}}}_o =$

$\alpha\phi(t, \bar{c}) + (1 - \alpha)\phi(t, \bar{\bar{c}}) = \dot{\tilde{x}} = \phi(c^{o})$. By concavity of ϕ, we know that $\phi(c^{o}) = \dot{\tilde{x}} \leq \phi(t, \tilde{c})$ and the answer is affirmative in view of the monotonicity of ϕ. This establishes proposition 4.

4. Pareto Optimality and Competitive Equilibrium in a General Equilibrium Model of Economic Growth

4.1 Introduction.

In this section we show that a growth program is "Pareto Optimal" if and only if it could be characterized by a "competitive equilibrium." The assumptions under which this is achieved are, essentially, non-increasing returns to scale concavity, of utility functions and independence on the consumption and production sides. We assume also that all decisions are taken at the beginning of the planning period, and that utility functionals are represented by integrals of instantaneous utility functions. Thus we restrict our attention to a very special class of economies, and so we shall not pretend that our result is usable in the planning of economic development. Our result, however, is a step forward from static welfare economics, see e.g. Karlin [30], as time enters the analysis in some sense. Our view of the next step, is where utility functionals don't assume a special form, and where an adaptive set up is the framework for decision making.

The result of this section is an extension of my paper with Professor Takayama [17], where we did not allow for capital accumulation. The price we pay for the presence of capital accumulation is in terms of concavity assumptions.

3.2 Formulation of the problem.

Consider an economy with n goods, m production processes, ℓ investment processes and B consumers. Let $g_{ij}(t, u_j)$ denote the outcome, in terms of good i of operating process j at level u_j at time t. We follow the convention that $g_{ij} \leq 0$ if good i is a net input to process j and $g_{ij} > 0$ if good i is a net output of process j. The presence of t as an argument of g_{ij} indicates the possibility of technological change, and the fact that g_{ij} depends only on x_j assumes the absence of externalities. Investment processes represent some grouping of investment projects (an alternative name would be: project types). They are described by way of requirement functions and addition-to-capacity functions.

Let $s_{ik}(t, v_k)$ be the requirement in terms of good i, when process k is operated at level v_k at time t, and let $S_{kj}(t, v_k)$ be the gross addition to the capacity of production process j when investment process k is operated at level v_k at time t. The capacity of a production process, $z_j(t)$, is defined to be the maximum level at which the process could be operated. Denote consumer β's consumption of good i at time t by c_i^β, and let $c_i = (c_i^1, \ldots, c_i^B)$. Assume all decision makers' planning period to be [0,T]. The rate of change in the capacity of production process j is given by:

(1) $$\dot{z}_j = \sum_k S_{kj}(t, v_k) - \alpha_j z_j$$

where α_j is the constant rate of depreciation in the capacity of process j.

Definition 1. A <u>program</u> is: vector functions: $u(t) = (u_1(r), \ldots, u_m(r))$, $v(t) = (v_1(r), \ldots, v_\ell(t))$, $c(t) = (c_1(t), \ldots, c_n(t))$; an initial capacity $z(0) = (z_1(r_o), \ldots, z_m(t_o)) = z_o$ and a terminal capacity $z(T) = z_T$. Let a program be denoted by $\Pi = (u, v, c, z, z_T)$.

Definition 2. A program is said to be <u>feasible</u> if:

(i) Equation (1) is satisfied subject to $z^o \in Z^o$ and $z^1 \in Z^1$, where Z^o and Z^1 are initial and terminal sets of capacities, with Z^1 representing the "goals" of the plan.

(ii) No more of any good is used than what is available.

(2) $$\sum_j g_{ij}(t, u_j) - \sum_k s_{ik}(t, v_k) - \sum_\beta c_i^\beta \geq 0$$

(iii) No process is operated beyond its capacity.

(3) $$u_j \leq z_j$$

(iv) Consumption and levels of processes are never negative.

(4) $$c_i \geq 0,\ v_j \geq 0,\ u_j \geq 0.$$

Now let $\phi^\beta(t, c^\beta)$ be the instantaneous utility function of consumer β, where $c^\beta = (c_1^\beta, \ldots, c_n^\beta)$. We assume that the consumers' ordering of consumption paths is represented by the functional

$$I^\beta[C^\beta] = \int_o^T {}^\beta(t, c^\beta)dt.$$

Def nition 3. A feasible program $\hat{\Pi}$ is said to be Pareto optimal if there does not exist a feasible program Π' such that $I^\beta[C'^\beta] \geq I^\beta[\hat{C}^\beta]$, $\beta = 1, \ldots, B$; with strict inequality for at least one β.

Before we introduce the definition of a competitive equilibrium path we first define the "cumulative return on a unit of capacity." Suppose capacity of type j depreciates at the rate α_j, then a unit of capacity added at time t becomes $e^{\alpha_j(t-\tau)}$ at time τ. If the rental per unit of capacity at time τ is $r_j(\tau)$, then the return accumulated from time t to time T is $\int_t^T e^{\alpha_j(t-\tau)} r_j(\tau)d\tau$.

Definition 4. A program $\overset{o}{\Pi}$ is said to be a competitive equilibrium program if there exist prices of goods $P \in E_n^+$, rentals of capacity $R \in E_m^+$, where P(t) and r(t) are piecewise continuous on [0,T], such that:

(i) Consumers maximize their instantaneous utility among bundles that cost no more, i.e. $u^\beta(c^\beta, t)$ is maximized at $\overset{o}{c}^\beta$ subject to $P.\overset{o}{C}^\beta \geq P.C^\beta$, at each point of time.

(ii) Producers of current goods maximize their profits, i.e. $\sum_i P_i g_{ij}(u_j, t) - r_j u_j$ is maximized at each point of time.

(iii) Producers of investment goods maximize their profits which is defined as the difference between cumulative returns on capacity added by investment minus the costs of production, i.e.,

$$\sum_j \left(\int_t^T e^{\alpha_j(t-\tau)} r_j(\tau)d\tau\right) S_{kj}(V_k, t) - \sum_i P_i S_{ij}(V_k, t) \text{ is maxim zed at each}$$

point of time.

(iv) Supply of current output is no less than demand; in case of excess supply of any good its price is zero, i.e.,

$$E_i = \sum_j y_{ij}(u_j, t) - \sum_\beta c_i^\beta - \sum_k S_{ik}(t, v_k) \geq 0 \text{ and } P_i E_i = 0.$$

(v) Utilized capacity is no more than available capacity; in case of excess cap city the rental is zero, i.e.,

$$z_j(t) - u_j(t) \geq 0 \text{ and } r_j(z_j(t) - u_j(t)) = 0.$$

Before we state our proposition, we list the assumptions under which it holds.

A.1 The functions g_{ij}, S_{kj}, s_{ik}, ϕ^β have continuous first order partial derivatives in all their arguments.

A.2 The functions g_{ij}, S_{kj}, $-s_{ik}$, ϕ^β are concave for any given t, positive g_{ij}. When $g_{ij} < 0$ g_{ij} is convex.

A.3 An increase in the level of an investment process never decreases the capacity of any production process. For each process j there exists an investment process k such that increasing the level of k adds to the capacity of j, i.e., S_{kj} is monotone nondecreasing, for each j there exists a k such that S_{kj} is strictly monotone increasing.

A.4 There exists a good that is a necessity in the sense that each consumer consumes a positive amount of that good; furthermore the "marginal utility" of that good is never zero, i.e., there exists a good i_o such that

(i) $c^{\beta}_{i_o}(t) > 0 \qquad t \in [0,T]$

(ii) $\phi^{\beta}_{i_o} = \partial \dfrac{\phi^{\beta}}{\partial c^{\beta}_{i_o}} \neq 0 \qquad t \in [0,T],\ c^{\beta}_{i_o} > 0.$

A.5 For each point in time, for any good i there exists a consumer that acquires a positive amount of that good, i.e., given $\bar{t} \in [0,T]$, and $i \in \{1, 2, \ldots, n\}$; $\bar{\beta} \in \{1, \ldots, B\} \cdot c^{\bar{\beta}}_{\underline{i}}(\bar{t}) > 0$.

A.6 The set Z_o is a single point, and the set Z_1 is an open set.

3.3 Characterization of the solution.

Proposition: Assuming A.1-A.6, a program $\hat{\Pi}$ is Pareto optimal if and only if it is a competitive equilibrium.

Proof:

(1) A program is $\hat{\Pi}$ Pareto optimal if it is a solution to all of the following maximization problems $M(\beta^*)$, $\beta^* = 1, 2, \ldots, B$, where $M(\beta^*)$ is:

$$\max I^{\overset{*}{\beta}}[c^{\overset{*}{\beta}}] \text{ subject to}$$

$$I^{\beta}[c^{\beta}] \geq I^{\beta}[\hat{c}^{\beta}],\ \beta \neq \overset{*}{\beta}$$

and to conditions (1) through (5).

To establish (1), we first suppose $\hat{\Pi}$ is not a solution to the individual maximization problems $M(\overset{*}{\beta})$. Then, for at least one $\overset{*}{\beta}$, there exists a feasible program $\bar{\Pi}$ such that

$$I^{\overset{*}{\beta}}[\bar{c}^{\overset{*}{\beta}}] > I^{\overset{*}{\beta}}[\hat{c}^{\overset{*}{\beta}}],$$

$$I^{\beta}[\bar{c}^{\beta}] > I^{\beta}[\hat{c}^{\beta}],\ \beta \neq \overset{*}{\beta}.$$

But this contradicts the Pareto optimality of $\hat{\Pi}$.

On the other hand, suppose $\acute{\Pi}$ is a solution to the problems $M(\beta^*)$ which is not a Pareto optimum. Then there exists a feasible program $\bar{\Pi}$ such that

$$I^\beta[\bar{c}^\beta] \geq I^\beta[c'^\beta],$$

with strict inequality for at least one β. Thus $\acute{\Pi}$ does not solve at least one of the problems $M(\beta^*)$, which is a contradiction. This completes the proof of (1).

(2) A program $\hat{\Pi}$ is a solution to a problem $M(\beta^*)$ if: there exist positive constants $y_\beta^{\beta*}$ and non-negative functions $\mu_i^{\beta*}(t)$, $\nu^{\beta*}(t)$ such that: for all $t \in [0,T]$,

2.1 $\mu_i^{\beta*}\hat{g}_{ij}' - \nu_j^{\beta*} \leq 0$, where $\hat{g}_{ij}' = \frac{\partial g_{ij}}{\partial u_j}\Big|\hat{p}$, with equality if $\hat{u}_j > 0$

2.2 $\sum_j(\int_t^T e^{\alpha_j(t-\tau)} \nu_j^{\beta*}(\tau)d\tau)\, \hat{S}_{kj}' - \sum_i \mu_i^{\beta*} \hat{s}_{ik}' \leq 0$, where

$$\hat{S}_{kj}' = \frac{\partial S_{kj}}{\partial v_k}\Big| , \quad \hat{s}_{ik}' = \frac{\partial s_{ik}}{\partial v_k}\Big| , \text{ with equality when } \hat{v}_k > 0.$$

2.3 $y_\beta^{\beta*}\hat{\phi}_i^\beta - \mu_i^{\beta*} \leq 0$, where $\hat{\phi}_i^\beta = \frac{\partial S_{ik}}{\partial v_k}\Big|$, with equality when $\hat{c}_i^\beta > 0$.

2.4 $\mu_i^{\beta*}(\sum_i g_{ij}(t, \hat{u}_j) - \sum_k s_{ik}(t, \hat{v}_k) - \sum_\beta \hat{c}_i^\beta) = 0$, $\nu_j^{\beta*}(\hat{z}_j - \hat{u}_j) = 0$.

To prove (2) we utilize the fact that each problem (β^*) is an optimal control problem. A necessary[1] (see theorem 1, this chapter) condition for $\hat{\Pi}$ to be a solution to the problem (β^*) is: there exist non-negative constants $y_\beta^{\beta*}$, functions $q_j^{\beta*}(t)$ and non-negative functions $\mu_k^{\beta*}(t)$, $\nu_j^{\beta*}(t)$, which are never all zeros, where $\mu_i^{\beta*}(t)$, $\nu_j^{\beta*}(t)$ are piecewise continuous and $q_j^\beta(t)$ are piecewise continuously differentiable; such that if we define:

$$H = y_\beta^{\beta*}\phi^{\beta*}(c^{\beta*}) + \sum_{\beta\neq\beta} y_\beta^{\beta*}(\phi^\beta(c^\beta) - \phi^\beta(\hat{c}^\beta)) + \sum_j q_j^{\beta*}(t)\,(\sum_k S_{kj}(t, v_k) - \alpha_j z_j)$$

$$+ \sum_i \mu^{\beta*}(\sum_j g_{ij}(t, u_j) - \sum_k s_{ik}(t, v_k) - \sum_\beta c_i^\beta) + \sum_j \nu_j^{\beta*}(z_j - u_j),$$

(i) $\dot{q}_j^{\beta*} = -\hat{H}_{z_j} = \frac{\partial H}{\partial z_j}\Big| = \alpha_j q_j^{\beta*} - \nu_j^{\beta*}$, $q(T) = 0$, $(t \in [0,T]$, all $j)$

---

[1] The regularity conditions for the theorem are satisfied (due to A.1), in particular the Jacobian of constraints 2-4 has full rank (it contains submatrix of order n+m whose determinant is 1).

(ii) $\hat{H}_{c_i^\beta} = \frac{\partial H}{\partial c_i^\beta} | = y_\beta^{\beta*} \hat{\phi}_i^\beta - \mu_i^{\beta*} \leq 0$ ($t \in [0,T]$, all i and all β)

(iii) $\hat{H}_{v_k} = \frac{\partial H}{\partial v_k} | = \sum_j q_j^{\beta*} \hat{S}'_{kj} - \sum_i \mu_i^{\beta*} \hat{s}'_{ik} \leq 0$ ($t \in [0,T]$, all k)

(iv) $\hat{H}_{u_j} = \frac{\partial H}{\partial u_j} | = \sum_i \mu_i^{\beta*} \hat{g}_{ij} - \nu_j^* \leq 0$ ($t \in [0,T]$, all j);

with equality holding if c_i, v_k, u_j are positive, respectively.

(v) $\mu_i^{\beta*}(\sum_j g_{ij}(t, \hat{u}_j) - \sum_k s_{ik}(t, \hat{v}_k) - \sum_\beta \hat{c}_i^\beta) = 0$, $\nu_j^{\beta*}(\hat{z}_j - \hat{u}_j) = 0$.

Solving (i) we find that $q_j^{\beta*}(t) = \int_t^T e^{\alpha_j(t-\tau)} \nu^{\beta*}(\tau)d\tau$ (utilizing $q_j^{\beta*}(T) = 0$ and the fact that (i) is a linear differential equation). Substituting in (iii)

(iii)' $\sum_j(\int_t^T e^{\alpha_j(t-\tau)} \nu_j^{\beta*}(\tau)d\tau) \hat{S}'_{kj} - \sum_i \mu_i^{\beta*} \hat{s}'_{ik} \leq 0$, ($t \in [0,T]$, all k).

Our necessary conditions are equivalent to (ii), (iii), (iv) and (v).

Note that $\int_t^T e^{\alpha_j(t-\tau)} \nu^\beta \geq 0$.

We now show that $y_\beta^{\beta*}$ are positive for all β. For, suppose one of them, say $y_{\bar{\beta}}^{\beta*}$, is zero. Then, by assumption A.4 (i) and by (ii) above we have

$$0 = y_{\bar{\beta}}^{\beta*} \hat{\phi}_{i_o} - \mu_{i_o}^{\beta*} = - \mu_{i_o}^{\beta*}, \text{ i.e. } \mu_{i_o}^{\beta*} = 0, \text{ while } 0 = y_\beta^{\beta*} \hat{\phi}_{i_o} - \mu_{i_j}^{\beta*}$$

$$= y_\beta^{\beta*} \hat{\phi}_{i_o} \ (\beta \neq \bar{\beta}).$$

By assumption A.4 (ii), this implies $y_\beta^{\beta*} = 0$ ($\beta \neq \bar{\beta}$) and thus $y_{\bar{\beta}}^{\beta*} = 0$ (all β). By A.5 (i) for each commodity i, there exists a consumer β such that $y_\beta^{\beta*} \hat{\phi}_i^\beta - \mu_i = 0$. Thus all the μ_i's are zeros. Substituting in (iii) we have $\sum_j q_j(t)\hat{S}'_{kj} \leq 0$. By A.3, $\hat{S}'_{kj} \geq 0$, we noted above that $q_j(t) \geq 0$. Thus $\sum_j q_j(t)\hat{S}'_{kj} = 0$, which (since $q_k S'_{kj} \geq 0$) implies $q_j(t)\hat{S}'_{kj} = 0$ for all k and j. By assumption A.3, for each j there exists a k such that $S'_{kj} > 0$. Thus $q_j(t) = 0$ for all j. This means $\int_t^T e^{\alpha_j(t-\tau)} \nu_j(\tau)d\tau \equiv 0$. But, since the integrand is non-negative, this implies $e^{\alpha_j(t-\tau)} \nu_j^{\beta*}(\tau) = 0$ a.e. for $\tau \in [t, T]$. But $e^{\alpha_j(t-\tau)} > 0$. Thus $\nu_j^{\beta*}(\tau) = 0$ a.e. for $\tau \in [t, T]$ for $\tau \in [0, T]$.

Thus $\nu_j^{\beta*}(t) = 0$ on a subset on $[0,T]$ of positive linear measure, say τ, for all j. Thus there exists a $t \in T$ for which $y_\beta^{\beta*}$, $u_i^{\beta*}$, $\nu_j^{\beta*}$ and $q_j^{\beta*}$ are simultaneously zeros. This contradiction proves the assertion that $y_\beta^{\beta*} > 0$. Conditions 2.1, 2.2, 2.3 and 2.4 are implied by as (ii), (iii), (iv) and (v) respectively.

Turning now to sufficiency, suppose conditions 2.1-2.4 hold. Then conditions (i) to (iv) are satisfied. By assumptions A.1 and A.2, the sufficiency theorem 2.a, this chapter, applies. Thus 2.1-2.3 are sufficient for $\hat{\Pi}$ to be a solution to problem $M(\beta*)$.

(3) If $\hat{\Pi}$ is a Pareto optimum then $\hat{\Pi}$ is a competitive equilibrium. Since $\hat{\Pi}$ is a Pareto optimum, we have, by (1) and (2): conditions 2.1-2.3 hold for $\beta* = 1, 2, \ldots, B$. Summing those sets of relations over $\beta*$, we get:

3.1 $$\left(\sum_{\beta*=1}^{B} \mu_i^{\beta*}\right)\hat{g}_{ij} - \left(\sum_{\beta*} \nu_j^{\beta*}\right) \leq 0, \text{ with equality if } \hat{u}_j > 0.$$

3.2 $$\sum_j \int_t^T e^{-\alpha_j(t-\tau)} \sum_{\beta*} \nu_j^{\beta*}(\tau)d\tau\, \hat{S}'_{kj} - \sum_i\left(\sum_{\beta*} \mu_i^{\beta*}\right) \hat{s}'_{ij} \leq 0, \text{ with equality if } \hat{v}_k > 0.$$

3.3 $$\left(\sum_{\beta*} y_\beta^{\beta*}\right) \hat{\phi}_i^\beta - \sum_{\beta*} \mu_i^{\beta*} \leq 0, \text{ with equality if } \hat{c}_i^\beta > 0.$$

3.4 $$\left(\sum_{\beta*} \mu_i^{\beta*}\right) \left(\sum_j g_{ij}(t, \hat{u}_j) - \sum_k s_{ik}(t, \hat{v}_k) - \sum_\beta \hat{c}_i^\beta\right) = 0, \quad \left(\sum_{\beta*} \nu_j^{\beta*}\right)(\hat{z}_j - \hat{u}_j) = 0.$$

Now define $p_i = \sum_{\beta*} \mu_i^{\beta*}$, $r_j = \sum_{\beta*} \nu_j^{\beta*}$, and $y^\beta = \sum_{\beta*} y_\beta^{\beta*}$. Substituting in 3.1-3.4, we get:

3.1' $$\sum_i p_i \hat{g}'_{ij} - r_j \leq 0.$$

3.2' $$\sum_j\left(\int_t^T e^{\alpha_j(t-\tau)} r_j(\tau)d\tau\right)\hat{S}'_{kj} - \sum_i p_i \hat{s}_{ij} \leq 0.$$

3.3' $$y^\beta \hat{\phi}_i^\beta - p_i \leq 0.$$

3.4' $$p_i\left(\sum_j g_{kj}(t, \hat{u}_j) - \sum_k s_{ik}(t, \hat{v}_k) - \sum_\beta c_i^\beta\right) = 0, \quad r_j(\hat{z}_j - \hat{u}_j) = 0.$$

Condition (i) in the definition of competitive equilibrium is satisfied by 3.3' and the concavity ϕ^β. For then we have, for $c_i^\beta \geq 0$,

$$y^{\beta}\phi^{\beta}(t, \hat{c}^{\beta}) - \sum_i p_i \hat{c}_i^{\beta} \geq y^{\beta}\phi^{\beta}(c_i^{\beta}) - \sum p_i c_i^{\beta}, \text{ where } y^{\beta} > 0.$$

This, in turn, implies:

$$\phi^{\beta}(c^{\beta}) \geq \phi^{\beta}(c^{\beta}), \text{ for } c_i^{\beta} \geq 0 \text{ with } \sum_i p_i c_i^{\beta} \leq \sum p_i \hat{c}_i^{\beta}.$$

Conditions (ii) and (iii) follow similarly from (3.1)' and (3.2)' respectively. The first parts of conditions (iv) and (v) follow from the feasibility of $\hat{\Pi}$. The second parts of conditions (iv) and (v) follow from (3.4)'.

(4) If $\hat{\Pi}$ is a competitive equilibrium then $\hat{\Pi}$ solves all maximization problems $M(\beta^*)$. Since $\hat{\Pi}$ is a competitive equilibrium, there exist functions, $p_i(t) \geq 0$, $r_j(t) \geq 0$, that are piecewise continuous such that:

4.1 $\hat{\phi}_i^{\beta} - p_i \leq 0$, with equality when $\hat{c}_i^{\beta} > 0$, (since $\phi^{\beta}(c^{\beta}) - \sum p_i c_i^{\beta}$ is maximized by (i), and such that (3.1)', (3.2)' and (3.4)' are satisfied (this follows from (ii), (iii), (iv) and (v) in the definition of competitive equilibrium). Now take $y_{\beta}^{\beta*} = 1$ for all β and β^*, $\mu_i^{\beta *} = p_i$ for all β^* and i, $\nu_j^{\beta*} = r_j$ for all j and β^* and $q_j^{\beta} = \int_t^T e^{\alpha_j(t-\tau)} r_j(\tau)d\tau$ for all j and β^*. Consider a problem $M(\overline{\beta}^*)$. With our choice of multipliers we have the conditions of (2) above. Thus by A.2, (2) applies. Since $\overline{\beta}^*$ is arbitrary this establishes (4).

(5) If $\hat{\Pi}$ is a competitive equilibrium then it is a Pareto optimum. This follows from (4) and (1). By (3) and (5) our proposition is proved.

5. Two Static Economic Examples:

The vast majority of applications of calculus of variations to economic analysis have been dynamic, in the sense of ch-osing among time paths. This was done by using criterion functionals that are integrals of instantaneous criterion functions, which involves a severe restriction on the form of such functionals. We shall present, hwere, two examples from economic analysis where the use of integrals is more natural. The problems are of some interest in themselves. The first problem deals with Ricardian rent and was first solved by Samuelson [46]. We shall provide a reformulation and proof of the results of the appendix of [46]. The second problem attempts to answer a question raised by Qayum [43] about investment maximizing tax schedules; whether they are progressive or regressive. We answer the question for the case considered by Qayum [43]. In general the answer depends, even in our very simple model, on such factors as the minimum amount of taxes to be collected and on the nature of the frequency distribution of income.

1. Ricardian Rent.

We shall use Samuelson's notation and formulation of the problem in [46]. The problem of determining the margin of expansion may be stated as follows:

$$\text{Maximize } 2\Pi \int_0^R f^1[x^o(\rho)L^1(\rho),\ L^1(\rho);\ \rho]\rho d\rho$$

Subject to:

1) $x - 2\Pi\int_0^R x^1(\rho)\rho d\rho \geq 0,$

2) $\Pi R^2 - 2\Pi\int_0^R L^1(\rho)\rho d\rho \geq 0,$

3) $L^1(\rho) \geq 0,\ x^1(\rho) \geq 0$

Proposition: If f is continuously differentiable then a necessary condition for $\hat{x}^1(\rho)$, $\hat{L}^1(\rho)$ and $\hat{R} > 0$ to be a solution to our problem is that there exist a land rental and a labor wage; $P_L > 0$ and $P_x > 0$ such that,

1) Marginal productivity of labor $\leq P_x$ with equality if $x^1(\rho)$ is positive, i.e.

(i) $\hat{f}_1\hat{L}^1(\rho) \leq P_x$, with equality if $\hat{x}(\rho) > 0$, for $0 < \rho \leq \hat{R}$.

2) Marginal productivity of land $\leq P_L$ with equality if $\hat{x}^1(\rho) > 0$, i.e.

(ii) $\hat{f}_1\hat{x}^1(\rho) + \hat{f}_2 \leq P_L$ with equality if $\hat{L}^1(\rho) > 0$, for $0 < \rho \leq \hat{R}$.

3) At the optimal frontier, the "rent" is at most zero, i.e.

(iii) $f^1[\hat{x}^1(\hat{R})\hat{L}(\hat{R}),\ \hat{L}^1(\hat{R});\ R] - P_x\hat{X}(R) - P_L\hat{L}(\hat{R}) \leq 0$, where rent is defined as the left hand side of (iii).

(iv) $P_x(x - 2\Pi \int_0^{\hat{R}} \hat{x}^1(\rho)\rho d\rho) = 0,\ x - 2\Pi \int_0^{\hat{R}} \hat{x}^1(\rho)\rho d\rho \geq 0.$

(v) $P_L(\Pi\hat{R}^2 - 2\Pi \int_0^{\hat{R}} \hat{L}(\rho)\rho d\rho) = 0,\ \Pi\hat{R}^2 - 2\Pi \int_0^{\hat{R}} L^1(\rho)\rho d\rho \geq 0.$

Proof: The proposition is established by applying theorem of after some transformations - theorem of .[1]

The necessary conditions are: there exist constants $\lambda_0, \lambda_1, \lambda_2$, non-negative and not all zero such that:

(1) $2\Pi\lambda_0 \hat{f}_1\hat{L}\rho - 2\Pi\lambda_1\rho \leq 0$, with equality is $\hat{x}(\rho) > 0;\ \rho \in [0, \hat{R}]$.

(2) $2\Pi\lambda_0(\hat{f}_1\hat{x} + \hat{f}_2)\rho - 2\Pi\lambda_2\rho \leq 0$, with equality if $\hat{L}(\rho) > 0;\ \rho \in [0, \hat{R}]$.

(3) $-[2\lambda_o\ \Pi\hat{R}\ \hat{f}[\hat{x}^1(\hat{R})\hat{L}^1(\hat{R}),\ \hat{L}^1(\hat{R});\ \hat{R}] - 2\lambda_1\ \Pi\hat{x}(\hat{R})\hat{R} - 2\lambda_2\Pi\hat{L}(\hat{R})\hat{R}] - 2\Pi\lambda_2\hat{R} = 0.$

(4) $\lambda_1(x - 2\Pi \int_0^{\hat{R}} \hat{x}^1(P)\rho d\rho) = 0.$

(5) $\lambda_2(\Pi\hat{R}^2 - 2\Pi \int_0^{\hat{R}} \hat{L}^1(\rho)\rho d\rho = 0.$

Clearly $\lambda_0 > 0$; for if it is zero then so would be λ_1 and λ_2 (by (1) and (2)) which would be a contradiction. Thus we may take $\lambda_0 = 1$. Take $P_x = \lambda_1$ and $P_1 = \lambda_2$. Then (i) and (ii) follow from (1) and (2) for $\rho > 0$. And (iv) and (v) follow from (4) and (5) and the fact that $\hat{x}^1$ and $\hat{L}^1$ satisfy the constraints.

(iii) follows (3) since we have (by (3)):

$$\hat{f} - P_x\hat{x}(\hat{R}) - P_L\hat{L}(\hat{R}) = -P_L \leq 0.$$

2. Investment maximizing tax policy[2]:

In [43] Qayum asks the following question: Assuming that investment depends only on the rate of income taxation, is an investment maximizing tax policy regressive or progressive? To answer the question Qayum restricts himself to the class of quadratic, in income, tax rates. Then he determines the parameters that lead to selecting one function from this class by finite dimensional maximization methods.

---

[1] Samuelson, in [46], cites Bliss [10] where there is no treatment of problems with inequality constraints.

[2] I am grateful to Professor Qayum for suggesting this problem.

However, since he takes income as a continuun, the problem of choosing the optimal tax schedule is a problem of selecting a function from an admissible class of functions defined on the range of taxable income, i.e. we must be able to choose the optimal form and parameters of the function. In fact, we shall show that if an "optimal"[3] tax schedule exists, for the Qayum problem, among the class of differentiable schedules then we get a completely different form of the schedule.

We formulate the problem as a simple problem in the calculus of variations and characterize the solution of the general problem. We then solve the problem where the investment function and the frequency distribution of income are those assumed by Qayum.

Let y denote a certain level of income, and suppose y_0 and y_1 are the lowest and highest levels of taxable income. Denote by Y, the interval $[y_0, y_1]$. Let f(y) be the frequency distribution of income. f is defined and is assumed to be differentiable and monotone decreasing on Y. Let R(y) denote the amount of taxing income at level y. The function R(y) will be the unknown in the present problem. We shall pick it from among the class of piecewise differentiable[4] functions defined on Y with values between zero and one, i.e.

(1) $R(t) \in [0, 1]$.

Let t(y) be the tax paid by an individual with income y. Then $t(y) = \int_{y_o}^{y} R(\eta)d\eta$.

We represent this by way of the following differential equation:

(2) $\dot{t}(y) = R(y),\ t(y_o) = 0,$

where the dot over a variable will always represent differentiation with respect to y. The taxes collected from all people with income y is f(y)t(y) and total tax collection, T, is given by the functional:

(3) $T[R] = \int_{y_o}^{y_1} t(y)f(y)dy.$

Let S(R(y)) denote the proportion that is invested out of an income y. Then investment out of an income y is yS(R) and the investment of all persons with given by yS(R)f(y). Thus total investment, I, is given by the functional:

---

[3] In the sense of investment maximization.
[4] The choice of this class is dictated by pedagogical considerations.

(4) $I = I[R] + \omega = \int_{y_o}^{y_1} yS(R)f(y)dy + \omega,$

where the constant ω represents investment out of non-taxable income.

The problem of maximizing investment subject to a lower bound, say γ, on tax collection is equivalent to maximizing $I[R]$, since ω is a constant, subject to constraint (1), (2) and

(5) $T[R] - \gamma \geq 0.$

The following conditions are necessary[5] and sufficient[6] for R to be a solution, assuming that S is a concave function of R and that the second derivative of S exists and is continuous.

There exist constants λ_o and β and functions $\lambda(y)$, $\mu_1(y)$ and $\mu_2(y)$ such that

(6.1) $\lambda_o \geq 0,\ \beta \geq 0,\ \beta(T[\hat{R}] - \gamma) = 0.$

(6.2) $\dot{\lambda} = -\beta f(y),\ \lambda(y_1) = 0.$

(6.3) $\mu_1 \geq 0,\ \mu_1(1 - \hat{R}) = 0.$

(6.4) $\mu_2 \geq 0,\ \mu_2\hat{R} = 0.$

(6.5) The vector $(\lambda_o, \lambda, \mu_1, \mu_2, \beta)$ is never a zero vector.

(6.6) $\lambda_o yfS'(\hat{R}) + \lambda(t) - \mu_1(t) + \mu_2(t) = 0.$

From (6.2) it follows that:

(7) $\lambda(y) = \beta\int_{y}^{y_1} f(\eta)\, d\eta,$

implying that $\lambda(y) \geq 0$ on Y.

(A) $\hat{R}(y) < 1$ for some subinterval of Y with positive length. We can then show that $\lambda_o > 0$ and hence may be taken equal to 1. This follows, by contradiction, from (6.6), (7) and (6.5). For if $\lambda_o = 0$ then, since $\mu_1 = 0$, $\lambda(y) = -\mu_2 \leq 0$ and hence $\lambda_2 = 0$. This by (7), implies $\beta = 0$ contradicting (6.5). Condition (6.6) may now be written:

(8) $yfS'(\hat{R}) = \mu_1(t) - \lambda(t) - \mu_2(t).$

Assuming that $\hat{R}$ is in (0, 1) and that $S' < 0$ and $S'' < 0$ it is necessary and sufficient for it to be a solution that:

(9.1) $-yfS'(\hat{R}) = \lambda(y).$

(9.2) $\dot{\lambda}(y) = -\beta f.$

---

[5] See theorem 1 of this chapter.

[6] See theorem 2 of this chapter

Differentiating both sides of (9.1) with respect to y we have:

(9.3) $\dot{\lambda} = -(fS' + y\dot{f}S' + yfS''\dot{R})$.

By (9.2) and (9.3) we have:

(10) $\dot{R}yS'' = \beta - S'(1 + y\,\dot{f}/f)$.

From (10) we see that, since $S'' \leq 0$, the sign of $\dot{R}$ is opposite to the sign of $\beta - S'(1 + y\,\dot{f}/f)$. In case $(1 + y\,\dot{f}/f) > 0$ the tax is regressive. This, however, occurs when either $\dot{f}/f$ is positive, which does not make sense[7], or when $\dot{f}/f < 0$ and and $y|\dot{f}/f| < 1$. If $(1 + y\,\dot{f}/f) < 0$ then the sign of $\dot{R}$ depends on the magnitude of β in relation to $|1 + y\,\dot{f}/f|$.

By way of illustration we analyze the following example, considered by Qayum Let $f = Cy^{-m}$, $m > 2$, $S = AR - BR^2$, $A \leq 2B$. Note that S is concave, and that our analyses apply.

By (7), $\lambda = \beta \int_y^{y_1} f(\eta)\, d\eta = C\beta \int_y^{y_1} \eta^{-m}$, i.e.

(11) $\lambda = \frac{C\beta}{1-m}(y_1^{1-m} - y^{1-m})$.

Assuming that $0 < \hat{R} < 1$, (9.1) applies and, by (11),

(12) $R = \frac{A}{2B} + \frac{\beta}{2(1-m)B}\left(\left(\frac{y}{y_1}\right)^{m-1} - 1\right)$.

Differentiating with respect to y, we find that $\dot{R} = \frac{(m-1)\beta}{2(1-m)B}\,\frac{y^{m-2}}{y_1^{m-1}} = -\frac{\beta}{2B}\,\frac{y^{m-2}}{y_1^{m-1}} \leq 0$.

Thus the optimal tax schedule is regressive for all levels of income, contrary to Qayum's assertion that there are cases where the tax is progressive. We may check this last result by (10); substituting for f, S' and S'' we have:

(13) $\dot{R}y(-2B) = \beta - (A - 2BR)(1 - m)$, i.e.

$$\dot{R} = \frac{-\beta}{2By} + \frac{(A - 2BR)}{2By}(1 - m)$$

$$\leq \frac{-\beta}{2By} + \frac{2B(1 - R)}{2By}(1 - m)$$

$$= \frac{-\beta}{2By} + (1 - R)(1 - m) < 0, \text{ since } m > 2 \text{ and } R < 1.$$

[7]And does not happen anyway.

"Now you have heard the story of Sordello."

Browning: Sordello

REFERENCES

1) Akhiezer, N. I.: The Calculus of Variations. English Translation: Blaisdel Publishing Company, New York 1962.

2) Arrow, K. and Hurwicz, L.: Reduction of Constrained Maxima to Saddle Point Problems. Proceedings of the Third Berkeley Symposium on Mathematical Statistics and Probability. University of California Press, July 1955.

3) Arrow, K., Hurwicz, L., and Uzawa, H.: Constraint Qualification in Maximization Problems. Naval Research Logistics Quarterly, Vol. 8 (1961), pp. 175-191.

4) Averbukh, V. I. and Smolyanov, O. G.: The Theory of Differentiation in Linear Topological Spaces. Uspekhi Mt. Nauk 22: 6 (1967), pp. 201-260; English Translation in Russian Math. Surveys 22: 6 (1967), pp. 201-258.

5) Averbukh, V. I. and Smolyanov, O. G.: The Various Definitions of the Derivative in Linear Topological Spaces. Uspekhi Mt. Nauk 23: 4 (1968), pp. 67-116. English Translation in Russian Math. Surveys 23: 4 (1968), pp. 67-113.

6) Berge, C.: Espaces Topologique, foncions multivoques, dunod, Paris 1959. English Translation: Topological Spaces, MacMillan, New York 1963.

7) Berkovitz, L.: Variational Problems of Control and Programing. Journal of Mathematical Analysis and Applications 3 (1961), pp. 145-169.

8) Bliss, G.: The Problem of Lagrange in the Calculus of Variations. American Journal of Mathematics 52 (1930), pp. 673-743.

9) Bliss, G.: Normality and Abnormality in the Calculus of Variations. Transactions of the American Mathematical Society, Vol. XLIII (1938), pp. 365-376.

10) Bliss, G.: Lectures on the Calculus of Variations. University of Chicago Press, Chicago 1946.

11) Blum, E.: Minimization of Functionals with Equality Constraints. SIAM Journal on Control, Vol. 3 (1965), pp. 299-317.

12) Brady, C.: The Minimum of a Function of Integrals in the Calculus of Variations. Contributions to the Calculus of Variations 1938-1941, University of Chicago Press, Chicago.

13) Bruno, M.: Fundamental Duality Relations in the Pure Theory of Capital and Growth. Review of Economic Studies 36: 1 (1969), pp. 39-54.

14) Caratheodory, C.: Variationsrechnung und partielle Differential gleichungen erster Ordnung. B. G. Teuber, Berlin 1935. English Translation: Calculus of Variations and Partial Differential Equations of the First Order, Volumes I and II, Holen-Day, San Francisco, California 1965 and 1967.

15) Cezari, L.: An Existence Theorem in Problems of Optimal Control. SIAM Journal on Control, Vol. 3 (1965), pp. 7-22.

16) Dubouskii, A. Ya. and Milyutin, A. A.: Extremum Problems in the Presence of Restrictions. Zh. Vychisl Fiz. 5 (1965), pp. 395-453. English Translation: U.S.S.R. Computational Mathematics and Math Physics 5 (1965), pp. 1-80.

17) El-Hodiri, M. and Takayama, A.: Price Implications of Pareto Optimality in a Dynamic General Equilibrium Model of Economic Growth. Research Paper No. 26, Research Papers in Theoretical and Applied Economics, Department of Economics, University of Kansas.

18) El'sgol, L. E.: Qualitative Methods in Mathematical Analysis. Moscow 1955. English Translation as Volume 12, Translations of Mathematical Monographs, American Mathematical Society. Providence, Rhode Island 1964.

19) Ewing, G.: Calculus of Variations with Applications. Norton & Company, New York 1969.

20) Fleming, W.: Functions of Several Variables. Addison Wesley, Reading, Massachusetts 1965.

21) Gapushkin, V. F.: On Critical Points of Functionals in Banach Spaces. Mat. Sbornik 64 (1964), pp. 589-617. English Translation in Three Papers on Optimization Theory, PEREC Report. Purdue University, Department of Electrical Engineering 1966.

22) Goldstine, Herman H.: Minimum Problems in the Functional Calculus. Bul. American Math. Society 46 (1940), pp. 142-149.

23) Guignard, M.: Generalized Kuhn-Tuker Conditions for Mathematical Programming Problems in a Banach Space. SIAM Journal on Control 7 (1969), pp. 232-241.

24) Halanay, A.: Optimal Control Systems with Time Lag. SIAM Journal on Control 6 (1968), pp. 215-234.

25) Hestenes, M.: On Variations and Optimal Control Theory. SIAM Journal on Control 3 (1965), pp. 23-48.

26) Hestenes, M.: Calculus of Variations and Optimal Control Theory. John Wiley & Sons, New York 1966.

27) Hurwicz, L.: Programming in Linear Spaces. In Arrow, Hurwicz and Uzawa (eds). Studies in Linear and Nonlinear Programming. Stanford University Press, Stanford, California 1958.

28) John, Fritz: Extremum Problems with Inequalities and Subsidiary Conditions. In K. O. Friedrick, O. E. Neugebaur and J. J. Stokes (eds). Studies and Essays: Courant Anniversary Volume. Interscience Publishers, New York, New York 1948.

29) Kantorovich, L. V. and Akilov, G. P.: Functional Analysis in Normal Linear Spaces. Moscow 1959. English Translation: Pergamon Press 1964.

30) Karlin, S.: Mathematical Methods and Theory in Games. Programming and Economics, Vol. 1. Addison Wesely, Reading Massachusetts 1959.

31) Karush, W.: Minima of Functions of Several Variables with Inequalities as Side Conditions. University of Chicago Master's Thesis 1939.

32) Kuhn, H. W. and Tucker, A. W.: Nonlinear Programming. In Proc. Second Berkeley Symposium on Math. Stat. and Probability, J. Neuman (ed.). University of California Press, Berkeley, California 1951.

33) Lee, E. and Markus, L.: Foundations of Optimal Control Theory. John Wiley & Sons, New York, New York 1967.

34) Liusternik, L. and Sobolev, V.: Elements of Functional Analysis. Moscow 1951. English Translation: Ungar Publishing Company, New York, New York 1961.

35) McShane, E.: Existence Theorems for the Problem of Balza in the Calculus of Variations. Duke Mathematical Journal 7 (1940), pp. 28-61.

36) McShane, E.: On Multipliers for Lagrange Problems. American Journal of Mathematics 61 (1939), pp. 809-819.

37) McShane, E. and Botts, A.: Real Analysis. D. Van Nostrand, Princeton, New Jersey 1954.

38) Mangasarian, O. L.: Sufficient Conditions for the Optimal Control of Nonlinear Systems. SIAM Journal on Control 4 (1966), pp. 139-152.

39) Neustadt, L.: An Abstract Variational Theory with Applications to a Broad Class of Variational Problems, I and II. SIAM Journal on Control 4 (1966), pp. 505-528 and 5 (1967), pp. 90-138.

40) Pars, L.: An Introduction to the Calculus of Variations. John Wiley & Sons, New York, New York 1962.

41) Pennisi, L. L.: An Indirect Sufficiency Proof for the Problem of Lagrange with Differential Inequalities as Added Side Conditions. Transactions of the American Math. Society 74 (1953), pp. 177-198.

42) Pontriagin, L. S., Boltyanskii, Gamkrelidge and Mishchenko: The Mathematical Theory of Optimal Processes. John Wiley & Sons, New York, New York 1962.

43) Qayum, A.: Investment Maximizing Tax Policy. Public Finance 3-4 (1963).

44) Ramsey, R.: A Mathematical Theory of Savings. Economic Journal 33 (1928), pp. 543-559.

45) Russak, I.: On Problems with Bounded State Variables. Journal of Optimization Theory and Applications 5: 2 (1970), pp. 114-157.

46) Samuelson, P. A.: A Modern Treatment of the Ricardian Economy: The Pricing of Goods and of Labor and Land Services. Quarterly Journal of Economics 1959, pp.

47) Vainberg, M. M.: Variational Methods for the Study of Nonlinear Operations. Moscow 1956. English Translation: Holden-Day, San Francisco, California 1964.

48) Valentine, Fredric: The Problem of Lagrange with Differential Inequalities as Added Side Condition. In Contributions to the Calculus of Variations 1933-1937. University of Chicago Press, Chicago, Illinois 1937.

49) Canon, M., Cullum, C. and Polak, E.: Theory of Optimal Control and Mathematical Programming. McGraw-Hill Book Company, New York 1970.

Lecture Notes in Operations Research and Mathematical Systems

Vol. 1: H. Bühlmann, H. Loeffel, E. Nievergelt, Einführung in die Theorie und Praxis der Entscheidung bei Unsicherheit. 2. Auflage, IV, 125 Seiten 4°. 1969. DM 12,–

Vol. 2: U. N. Bhat, A Study of the Queueing Systems M/G/1 and GI/M/1. VIII, 78 pages. 4°. 1968. DM 8,80

Vol. 3: A. Strauss, An Introduction to Optimal Control Theory. VI, 153 pages. 4°. 1968. DM 14,–

Vol. 4: Einführung in die Methode Branch and Bound. Herausgegeben von F. Weinberg. VIII, 159 Seiten. 4°. 1968. DM 14,–

Vol. 5: L. Hyvärinen, Information Theory for Systems Engineers. VIII, 205 pages. 4°. 1968. DM 15,20

Vol. 6: H. P. Künzi, O. Müller, E. Nievergelt, Einführungskursus in die dynamische Programmierung. IV, 103 Seiten. 4°. 1968. DM 9,–

Vol. 7: W. Popp, Einführung in die Theorie der Lagerhaltung. VI, 173 Seiten. 4°. 1968. DM 14,80

Vol. 8: J. Teghem, J. Loris-Teghem, J. P. Lambotte, Modèles d'Attente M/G/1 et GI/M/1 à Arrivées et Services en Groupes. IV, 53 pages. 4°. 1969. DM 6,–

Vol. 9: E. Schultze, Einführung in die mathematischen Grundlagen der Informationstheorie. VI, 116 Seiten. 4°. 1969. DM 10,–

Vol. 10: D. Hochstädter, Stochastische Lagerhaltungsmodelle. VI, 269 Seiten. 4°. 1969. DM 18,–

Vol. 11/12: Mathematical Systems Theory and Economics. Edited by H. W. Kuhn and G. P. Szegö. VIII, IV, 486 pages. 4°. 1969. DM 34,–

Vol. 13: Heuristische Planungsmethoden. Herausgegeben von F. Weinberg und C. A. Zehnder. II, 93 Seiten. 4°. 1969. DM 8,–

Vol. 14: Computing Methods in Optimization Problems. Edited by A. V. Balakrishnan. V, 191 pages. 4°. 1969. DM 14,–

Vol. 15: Economic Models, Estimation and Risk Programming: Essays in Honor of Gerhard Tintner. Edited by K. A. Fox, G. V. L. Narasimham and J. K. Sengupta. VIII, 461 pages. 4°. 1969. DM 24,–

Vol. 16: H. P. Künzi und W. Oettli, Nichtlineare Optimierung: Neuere Verfahren, Bibliographie. IV, 180 Seiten. 4°. 1969. DM 12,–

Vol. 17: H. Bauer und K. Neumann, Berechnung optimaler Steuerungen, Maximumprinzip und dynamische Optimierung. VIII, 188 Seiten. 4°. 1969. DM 14,–

Vol. 18: M. Wolff, Optimale Instandhaltungspolitiken in einfachen Systemen. V, 143 Seiten. 4°. 1970. DM 12,–

Vol. 19: L. Hyvärinen, Mathematical Modeling for Industrial Processes. VI, 122 pages. 4°. 1970. DM 10,–

Vol. 20: G. Uebe, Optimale Fahrpläne. IX, 161 Seiten. 4°. 1970. DM 12,–

Vol. 21: Th. Liebling, Graphentheorie in Planungs- und Tourenproblemen am Beispiel des städtischen Straßendienstes. IX, 118 Seiten. 4°. 1970. DM 12,–

Vol. 22: W. Eichhorn, Theorie der homogenen Produktionsfunktion. VIII, 119 Seiten. 4°. 1970. DM 12,–

Vol. 23: A. Ghosal, Some Aspects of Queueing and Storage Systems. IV, 93 pages. 4°. 1970. DM 10,–

Vol. 24: Feichtinger, Lernprozesse in stochastischen Automaten. V, 66 Seiten. 4°. 1970. DM 6,–

Vol. 25: R. Henn und O. Opitz, Konsum- und Produktionstheorie I. II, 124 Seiten. 4°. 1970. DM 10,–

Vol. 26: D. Hochstädter und G. Uebe, Ökonometrische Methoden. XII, 250 Seiten. 4°. 1970. DM 18,–

Bitte wenden/Continued

Vol. 27: I. H. Mufti, Computational Methods in Optimal Control Problems. IV, 45 pages. 4°. 1970. DM 6,–.

Vol. 28: Theoretical Approaches to Non-Numerical Problem Solving. Edited by R. B. Banerji and M. D. Mesarovic. VI, 466 pages. 4°. 1970. DM 24,–

Vol. 29: S. E. Elmaghraby, Some Network Models in Management Science. III, 177 pages. 4°. 1970. DM 16,–

Vol. 30: H. Noltemeier, Sensitivitätsanalyse bei diskreten linearen Optimierungsproblemen. VI, 102 Seiten. 4°. 1970. DM 10,–

Vol. 31: M. Kühlmeyer, Die nichtzentrale t-Verteilung. II, 106 Seiten. 4°. 1970. DM 10,–

Vol. 32: F. Bartholomes und G. Hotz, Homomorphismen und Reduktionen linearer Sprachen. XII, 143 Seiten. 4°. 1970. DM 14,–

Vol. 33: K. Hinderer, Foundations of Non-stationary Dynamic Programming with Discrete Time Parameter. VI, 160 pages. 4°. 1970. DM 16,–

Vol. 34: H. Störmer, Semi-Markoff-Prozesse mit endlich vielen Zuständen. Theorie und Anwendungen. VII, 128 Seiten. 4°. 1970. DM 12,–

Vol. 35: F. Ferschl, Markovketten. VI, 168 Seiten. 4°. 1970. DM 14,–

Vol. 36: M. P. J. Magill, On a General Economic Theory of Motion. VI, 95 pages. 4°. 1970. DM 10,–

Vol. 37: H. Müller-Merbach, On Round-Off Errors in Linear Programming. VI, 48 pages. 4°. 1970. DM 10,–

Vol. 38: Statistische Methoden I, herausgegeben von E. Walter. VIII, 338 Seiten. 4°. 1970. DM 22,–

Vol. 39: Statistische Methoden II, herausgegeben von E. Walter. IV, 155 Seiten. 4°. 1970. DM 14,–

Vol. 40: H. Drygas, The Coordinate-Free Approach to Gauss-Markov Estimation. VIII, 113 pages. 4°. 1970. DM 12,–

Vol. 41: U. Ueing, Zwei Lösungsmethoden für nichtkonvexe Programmierungsprobleme. IV, 92 Seiten. 4°. 1971. DM 16,–

Vol. 42: A.V. Balakrishnan, Introduction to Optimization Theory in a Hilbert Space. IV, 153 pages. 4°. 1971. DM 16,–

Vol. 43: J. A. Morales, Bayesian Full Information Structural Analysis. VI, 154 pages. 4°. 1971. DM 16,–

Vol. 44: G. Feichtinger, Stochastische Modelle demographischer Prozesse. XIII, 404 pages. 4°. 1971. DM 28,–

Vol. 45: K. Wendler, Hauptaustauschschritte (Principal Pivoting). II, 64 pages. 4°. 1971. DM 16,–

Vol. 46: C. Boucher, Leçons sur la théorie des automates mathématiques. VIII, 193 pages. 4°. 1971. DM 18,–

Vol. 47: H. A. Nour Eldin, Optimierung linearer Regelsysteme mit quadratischer Zielfunktion. VIII, 163 pages. 4°. 1971. DM 16,–

Vol. 48: M. Constam, Fortran für Anfänger. VI, 143 pages. 4°. 1971. DM 16,–

Vol. 49: Ch. Schneeweiß, Regelungstechnische stochastische Optimierungsverfahren. XI, 254 pages. 4°. 1971. DM 22,–

Vol. 50: Unternehmensforschung Heute – Übersichtsvorträge der Züricher Tagung von SVOR und DGU, September 1970. Herausgegeben von M. Beckmann. IV, 133 pages. 4°. 1971. DM 16,–

Vol. 51: Digitale Simulation. Herausgegeben von K. Bauknecht und W. Nef. IV, 207 pages. 4°. 1971. DM 18,–

Vol. 52: Invariant Imbedding. Proceedings of the Summer Workshop on Invariant Imbedding Held at the University of Southern California, June – August 1970. Edited by R. E. Bellman and E. D. Denman. IV, 148 pages. 4°. 1971. DM 16,–

Vol. 53: J. Rosenmüller, Kooperative Spiele und Märkte. IV, 152 pages. 4°. 1971. DM 16,–

Vol. 54: C. C. von Weizsäcker, Steady State Capital Theory. III, 102 pages. 4°. 1971. DM 16,–

Vol. 55: P. A. V. B. Swamy, Statistical Inference in Random Coefficient Regression Models. VIII, 209 pages. 4°. 1971. DM 20,–

Vol. 56: Mohamed A. El-Hodiri, Constrained Extrema. Introduction to the Differentiable Case with Economic Applications. III, 130 pages. 4 . 1971. DM 16,–